AF361691

Optical In-Process Measurement Systems

Optical In-Process Measurement Systems

Editor

Andreas Fischer

MDPI • Basel • Beijing • Wuhan • Barcelona • Belgrade • Manchester • Tokyo • Cluj • Tianjin

Editor
Andreas Fischer
University of Bremen
Germany

Editorial Office
MDPI
St. Alban-Anlage 66
4052 Basel, Switzerland

This is a reprint of articles from the Special Issue published online in the open access journal *Applied Sciences* (ISSN 2076-3417) (available at: https://www.mdpi.com/journal/applsci/special_issues/Optical_Measurement).

For citation purposes, cite each article independently as indicated on the article page online and as indicated below:

LastName, A.A.; LastName, B.B.; LastName, C.C. Article Title. *Journal Name* **Year**, *Volume Number*, Page Range.

ISBN 978-3-0365-3849-5 (Hbk)
ISBN 978-3-0365-3850-1 (PDF)

© 2022 by the authors. Articles in this book are Open Access and distributed under the Creative Commons Attribution (CC BY) license, which allows users to download, copy and build upon published articles, as long as the author and publisher are properly credited, which ensures maximum dissemination and a wider impact of our publications.

The book as a whole is distributed by MDPI under the terms and conditions of the Creative Commons license CC BY-NC-ND.

Contents

About the Editor

Andreas Fischer is Full Professor at the Department of Production Engineering, University of Bremen, in Germany and is head of the Bremen Institute for Metrology, Automation and Quality Science (BIMAQ). He studied electrical engineering and completed his PhD at the Technische Universität Dresden. As one of several awards, he received the Measurement Technique Prize 2010 of the Society of University Professors of Measurement Technique in Germany. In 2021, he further received a prestigious ERC Consolidaor Grant for enabling indirect optical geometry measurements. His research interests include optical measurement principles for flow and production metrology, their near-process or in-process application, and the information theoretic analysis of achievable measurement uncertainty to surpass the current limits of measurability.

Preface to "Optical In-Process Measurement Systems"

To measure the means to gain information, information about technical processes is crucial. It is crucial for establishing sustainable and environmentally friendly processes toward improving our lives and society. The idea for the present book was born from the desire to highlight the respective ongoing attempts to transfer optical measurement principles from laboratories to real-world applications. Thanks to the support of many research groups worldwide, the book contains different measurement approaches for different technical processes covering the production and machining of parts as well as the use of renewable energy and further important topics. As a result, the book illustrates the versatility and the beauty of optical principles for enabling in-process measurements. I do hope that readers will be inspired when reading the fascinating novel works, and that they maybe trigger new ideas on how to exploit the full potential of optical measurements.

Andreas Fischer
Editor

applied sciences

Editorial

Special Issue on Optical In-Process Measurement Systems

Andreas Fischer

Bremen Institute for Metrology, Automation and Quality Science, University of Bremen, Linzer Str. 13, 28357 Bremen, Germany; andreas.fischer@bimaq.de

Citation: Fischer, A. Special Issue on Optical In-Process Measurement Systems. *Appl. Sci.* **2022**, *12*, 2664. https://doi.org/10.3390/app12052664

Received: 2 March 2022
Accepted: 3 March 2022
Published: 4 March 2022

Publisher's Note: MDPI stays neutral with regard to jurisdictional claims in published maps and institutional affiliations.

Copyright: © 2022 by the author. Licensee MDPI, Basel, Switzerland. This article is an open access article distributed under the terms and conditions of the Creative Commons Attribution (CC BY) license (https://creativecommons.org/licenses/by/4.0/).

1. Introduction

Optical principles enable precise measurements, down to the quantum mechanical limits, and provide the fastest possible measurement speed: the speed of light. Driven by the ongoing advances in powerful light sources, accurate light modulation possibilities, and efficient light detectors, the capabilities of optical measurement systems are increasing. However, a current challenge is to make use of the benefits of optical principles for in-process measurements on real-world objects. Examples are flow processes on wind turbines, on airplanes, and in combustors; the thermal and mechanical processes on a workpiece during manufacturing; and the exploration of natural processes on Earth and in space. Studying non-idealized, non-scaled objects during their actual operation is an important task for the field of measurement science, since it allows us to gain new insights from the *actual* process behavior to be engineered. Furthermore, in-process measurements are required to create in-process controls.

In-process measurement conditions are often challenging and can result in limited optical access, an uncooperative measurement environment, a large measurement distance, a large measurement object, or a low signal-to-noise ratio. In order to break new ground for the transition "from the lab to the app" (from the laboratory to the application), optical in-process measurements have to be realized at the limits of measurability and beyond.

2. Progress of Optical In-Process Measurement Systems

This Special Issue covers a wide range of different optical measurement principles and in-process applications.

Regarding production engineering, an overview of recent advances in the optical measurement of the geometry of manufactured components is presented in [1]. In addition, for in situ measurements of the pitch of a diffraction grating, the diffraction pattern of a mode-locked femtosecond laser is studied in [2], while a real-time automated system to perform dual-aperture common-path interferometer phase-shifting is proposed in [3]. Furthermore, an improved Mirau-type coherence scanning interferometer with vibration compensation is described in [4], which widens the range of industrial applicability.

With a focus on 100% inspection during laser material processing, an inline measurement approach based on low coherence interferometry is presented in [5]. Considering hairpin welding, a camera-based spatter detection method to assess the quality of the in-process weld is further investigated in [6]. Finally, the in-process measurement capability of speckle photography to quantify not only the 2D but the complete 3D deformation of a loaded workpiece is proved and characterized in [7].

Regarding the engineering of fluid flows, spectroscopy is applied to study the diffusional behavior of new insulating gas mixtures as alternatives to the use of SF6 in medium voltage switchgear in [8]. In addition, for the future development of an active explosion suppression system in coal mines, the optimal wavelength range for the optical sensing of the ignition of methane–air mixtures and coal dust is identified in [9].

Last but not least, two works concerning the in-process flow analysis for and on wind turbines are presented. To reliably assess the wind velocity and thus the suitability of a site for building a wind turbine, a novel transportable wind Lidar system with superior resolution is tested and successfully validated in [10]. Regarding the contactless investigation of premature flow transitions (from laminar to turbulent) on running wind turbines, a thermographic flow visualization with an automated, model-based detection is described in [11].

3. Outlook

This Special Issue highlights some of the recent achievements in advancing optical measurement systems for in-process applications in the fields of production engineering, process engineering and renewable energy systems. Since measurements are a key element for improving technical processes in terms of sustainability, efficiency and environment friendliness, optical in-process measurements are of increasing importance and will further shape our future world and society.

Funding: This research received no external funding.

Informed Consent Statement: Not applicable.

Acknowledgments: This issue would not be possible without the contributions of the various talented authors, hardworking and professional reviewers, and dedicated editorial team of *Applied Sciences*. To all of them, many thanks for your contribution and support.

Conflicts of Interest: The author declares no conflict of interest.

References

1. Bergmann, R.B.; Kalms, M.; Falldorf, C. Optical In-Process Measurement: Concepts for Precise, Fast and Robust Optical Metrology for Complex Measurement Situations. *Appl. Sci.* **2021**, *11*, 10533. [CrossRef]
2. Shin, D.W.; Quan, L.; Shimizu, Y.; Matsukuma, H.; Cai, Y.; Manske, E.; Gao, W. In-Situ Evaluation of the Pitch of a Reflective-Type Scale Grating by Using a Mode-Locked Femtosecond Laser. *Appl. Sci.* **2021**, *11*, 8028. [CrossRef]
3. Barcelata-Pinzón, A.; Álvarez-Tamayo, R.I.; Prieto-Cortés, P. A Real-Time Automated System for Dual-Aperture Common-Path Interferometer Phase-Shifting. *Appl. Sci.* **2021**, *11*, 7438. [CrossRef]
4. Serbes, H.; Gollor, P.; Hagemeier, S.; Lehmann, P. Mirau-Based CSI with Oscillating Reference Mirror for Vibration Compensation in In-Process Applications. *Appl. Sci.* **2021**, *11*, 9642. [CrossRef]
5. Zechel, F.; Jasovski, J.; Schmitt, R.H. Dynamic, Adaptive Inline Process Monitoring for Laser Material Processing by Means of Low Coherence Interferometry. *Appl. Sci.* **2021**, *11*, 7556. [CrossRef]
6. Hartung, J.; Jahn, A.; Bocksrocker, O.; Heizmann, M. Camera-Based In-Process Quality Measurement of Hairpin Welding. *Appl. Sci.* **2021**, *11*, 10375. [CrossRef]
7. Tausendfreund, A.; Stöbener, D.; Fischer, A. In-Process Measurement of Three-Dimensional Deformations Based on Speckle Photography. *Appl. Sci.* **2021**, *11*, 4981. [CrossRef]
8. Espinazo, A.; Lombraña, J.I.; Asua, E.; Pereda-Ayo, B.; Alonso, M.L.; Alonso, R.M.; Cayero, L.; Izcara, J.; Izagirre, J. Diffusional Behavior of New Insulating Gas Mixtures as Alternatives to the SF6-Use in Medium Voltage Switchgear. *Appl. Sci.* **2022**, *12*, 1436. [CrossRef]
9. Khokhlov, S.; Abiev, Z.; Makkoev, V. The Choice of Optical Flame Detectors for Automatic Explosion Containment Systems Based on the Results of Explosion Radiation Analysis of Methane- and Dust-Air Mixtures. *Appl. Sci.* **2022**, *12*, 1515. [CrossRef]
10. Wilhelm, P.; Eggert, M.; Hornig, J.; Oertel, S. High Spatial and Temporal Resolution Bistatic Wind Lidar. *Appl. Sci.* **2021**, *11*, 7602. [CrossRef]
11. Parrey, A.-M.; Gleichauf, D.; Sorg, M.; Fischer, A. Automated Detection of Premature Flow Transitions on Wind Turbine Blades Using Model-Based Algorithms. *Appl. Sci.* **2021**, *11*, 8700. [CrossRef]

Article

Optical In-Process Measurement: Concepts for Precise, Fast and Robust Optical Metrology for Complex Measurement Situations

Ralf B. Bergmann [1,2,3,*], Michael Kalms [1] and Claas Falldorf [1]

[1] Bremer Institut für Angewandte Strahltechnik GmbH (BIAS), Klagenfurter Str. 5, 28359 Bremen, Germany; kalms@bias.de (M.K.); falldorf@bias.de (C.F.)

[2] Angewandte Optik, Fachbereich Physik/Elektrotechnik, Universität Bremen, Otto Hahn Allee NW1, 28359 Bremen, Germany

[3] MAPEX, Universität Bremen, Bibliothekstr. 1, 28359 Bremen, Germany

* Correspondence: bergmann@bias.de

Abstract: Optical metrology is a key element for many areas of modern production. Preferably, measurements should take place within the production line (in-process) and keep pace with production speed, even if the parts have a complex geometry or are difficult to access. The challenge for modern optical in-process measurements is, therefore, how to simultaneously make optical metrology precise, fast, robust and capable of handling geometrical complexity. The potential of individual techniques to achieve these demands can be visualized by the tetrahedron of optical metrology. Depending on the application, techniques based on interferometry or geometrical optics may have to be preferred. The paper emphasizes complexity and robustness as prime areas of improvement. Concerning interferometric techniques, we report on fast acquisition as used in holography, tailoring of coherence properties and use of Multiple simultaneous Viewing direction holography (MultiView), self reference used in Computational Shear Interferometry (CoSI) and the simultaneous use of several light sources in Multiple Aperture Shear Interferometry (MArS) based on CoSI as these techniques have proven to be particularly effective. The use of advanced approaches based on CoSI requires a transition of the description of light from the use of the well-known wave field to the coherence function of light. Techniques based on geometric optics are generally comparatively robust against environmental disturbances, and Fringe Projection (FP) is shown to be especially useful in very demanding measurement conditions.

Keywords: in-process measurement; in situ measurement; optical metrology; quality control; interferometry; fringe projection; computational shear interferometry; coherence function; structure function; additive manufacturing

Citation: Bergmann, R.B.; Kalms, M.; Falldorf, C. Optical In-Process Measurement: Concepts for Precise, Fast and Robust Optical Metrology for Complex Measurement Situations. *Appl. Sci.* **2021**, *11*, 10533. https://doi.org/10.3390/app112210533

Academic Editor: Chien-Hung Liu

Received: 20 September 2021
Accepted: 31 October 2021
Published: 9 November 2021

Publisher's Note: MDPI stays neutral with regard to jurisdictional claims in published maps and institutional affiliations.

Copyright: © 2021 by the authors. Licensee MDPI, Basel, Switzerland. This article is an open access article distributed under the terms and conditions of the Creative Commons Attribution (CC BY) license (https://creativecommons.org/licenses/by/4.0/).

1. Introduction

Optical metrology is an important part of modern production [1,2], since light can be most easily and quickly tailored according to measurement requirements, is contact-free and can usually be employed as non-damaging. Concerning the setting of the measurement with respect to the production environment, there seems to be no generally accepted terminology on the definition of terms such as in-process or in situ metrology [3] and the like [4]. Recently, Gao et al. gave an extensive overview on surface metrology for precision manufacturing [1]. In their paper, the authors classify the related terms "so that they can be used distinguishably and properly [...] in the manufacturing community". The following items show their classification:

- In situ surface metrology: Measurement of the workpiece surface is carried out on the same work floor and in the same manufacturing environment, without isolating the workpiece from the manufacturing environment.

- In-line/On-line surface metrology: Measurement of the workpiece surface is carried out on a production line without moving the workpiece outside the production line.
- On-machine surface metrology: Measurement of the workpiece surface is carried out on a manufacturing machine where the workpiece is manufactured.
- In-process surface metrology: The on-machine measurement of the workpiece surface is carried out while the manufacturing process is taking place.

The classification given above will be used in the present paper as well. The authors of [1] further distinguish » "pre-process" on-machine surface metrology « conducted before the manufacturing process and » "post-process" on-machine surface metrology « conducted after the manufacturing process. These terms, especially the last one, will not be used here, since it is easily confused with the common understanding of [3].

- Post-process measurement: The measurement occurs after the finished workpiece is removed from the machine. A typical scenario is a measurement using a coordinate measuring machine to inspect the finished workpiece. Post-process measurements usually can afford time-consuming procedures and dedicated inspection areas, e.g., with vibration isolation, constant temperature or the like and is thus often used for high precision metrology that is not applicable in a production environment.

The challenge for modern optical metrology is how to create techniques that are simultaneously precise, fast, robust and capable of handling geometrical complexity in order to be useful as in-process or in situ measurement. Figure 1 shows the "Tetrahedron of Optical Metrology" for the characterization of measurement tasks and measurement systems [5]. It describes the measurement speed (S), the measurement uncertainty, named precision (P) here for short, the tolerance with respect to disturbances, such as mechanical vibrations, described as robustness (R), and the influence of the geometrical properties of the measurement object or the measurement situation, described as geometrical complexity (C). For every combination of S, P, R and C, there is a characteristic filling of the tetrahedron. Most measurement systems have their strengths in one or two of these properties, but the challenge lies in the combination of these properties.

Requirements for measuring systems result from the respective measuring task. For example, the measurement of a component with a component tolerance of 0.5 μm requires a measurement uncertainty of $\leq$50 nm, the scanning of an area of 1 cm^2 in steps of 1 μm in 1 s requires a data rate of 100 million points/s, and with a vibration amplitude of approximately 2 mm at 10 Hz, there is an acceleration of approximately 1 g. The geometric complexity of the object increases from spheres to aspheres to freeform surfaces deviating strongly from aspheres up to complex free forms, which may contain poorly accessible areas, holes or cavities.

Within Figure 1a, the small darker volume shows a typical situation for conventional interferometry in relation to the parameters of the tetrahedron. Optical metrology based on interferometrical methods [6] is usually very precise, but (intrinsically) also sensitive to mechanical disturbances. Vibrations of the object or optical components of the measurement setup can change the phase relationship of the reference and object beam to one another and may thus corrupt the measurement [7]. An overview of harsh measurement conditions that require robust metrology approaches is given in Reference [8]. In addition, conventional interferometry tends to have difficulties measuring geometrically complex objects as will be dealt with in this paper.

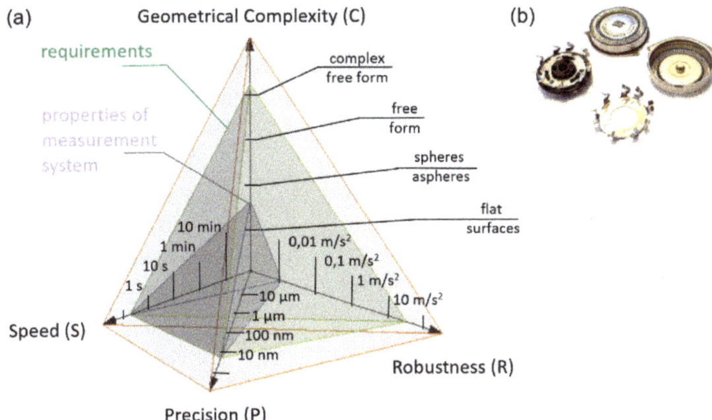

Figure 1. Tetrahedron of optical metrology and its applications: (**a**) Requirements are speed (S), robustness (R), precision (P) and geometric complexity (C) as described in the text. As an example, the small, darker shaded volume shows a measurement system, which may represent a given type of interferometer with high measurement speed, high precision but low robustness suitable for only simple geometric shapes such as flat or near flat surfaces. In contrast, a production process for certain micro-parts may not only require high speed and high precision but may also require a measurement process that allows measuring a complex geometry in the presence of high levels of vibrations as indicated in this scheme by the larger lighter shaded volume. (**b**) Example of complex shaped parts of a commercially available razor. The cap (upper part) has a diameter of approximately one cm. The ensemble consists of three parts with the cap also shown from the inside (on the right). The cap and the knife blades (upper and lower part) require suitable tolerances to guarantee proper cutting function.

There is a great variety of techniques for optical metrology. Fundamentals and principles are, for example, described in Reference [6,9–12], practical applications are outlined, for example, in References [13–16]. The size of measurement objects in most cases ranges from sub-µm to several m [2,13,17], in some cases even to the km-range [18]. It is, however, not the purpose of this paper to give a general overview on optical metrology but to demonstrate approaches that show a high potential to boost the performance of optical metrology for in-process measurements.

2. Methods and Scope of Approaches for In-Process Optical Metrology

In this paper, we emphasize two principal approaches for measuring shape, position or surface properties of objects: Geometric metrology using incoherent light is described by the laws of ray optics and uses structured light, mainly in the form of dynamic stripe patterns. Measurement techniques based on interferometry [19] are described by wave optics and use interference of light and the resulting fringe patterns. In general, the light is divided into two partial beams, one of which serves as a reference beam and the other as a measurement beam that interacts with the measurement object and thereby changes its phase distribution. If both partial beams are allowed to interfere with each other, the shape of the measurement object can be determined with very high accuracy, i.e., with a very low measurement uncertainty down to the nm range. More rugged approaches use differential techniques, which will be described later. It should be noted that interferometry is not necessarily associated with the use of laser sources. For a comparison of the methods, see, e.g., Reference [20].

One of the crucial requirements of in-line or even in-process inspection is the immunity towards mechanical distortions, which are unavoidable in a production environment. For metrology based on interferometry, this means a significant challenge. Interferometry is usually employed if high precision is required. However, vibrations with amplitudes 10 or 100 times larger than the measurement uncertainty shall not disturb the measurement. One solution is one-shot approaches with short exposure times, as can be employed in the case of digital holography. Another elegant solution to this problem is differential methods, which are intrinsically immune to vibrations.

At BIAS, we are concentrating on the interferometry-based techniques of Digital Holography (DH) and Computational Shear Interferometry (CoSI) and the geometrically based techniques of Deflectometry and Fringe Projection (FP). The following sections provide examples of measurements on small, medium and large scale objects and demonstrate, with respect to the scope of this paper, the development of robust, precise and fast measurement systems for complex measurement situations.

3. Selected Results of Optical Metrology Techniques

3.1. Techniques Based on Interferometry

3.1.1. Rapid Holographic Shape Measurement of Micro Parts

A 100% quality inspection of metallic micro components often fails due to the compromise between precision and speed. We therefore developed a fast measurement detection and quality inspection of micro components [21]. The system is based on digital holography (DH), an interferometric technique that allows to extract shape data from one or more recorded holograms [22].

The great advantage of DH is that the required holograms can be recorded within a fraction of a second, while it allows measuring with interferometric precision [11]. Additionally, the focus can be adjusted subsequent to the measurement, which allows for an extended depth of focus without the requirement of an optical scanning process. Finally, because digital holographic shape measurement is based on the evaluation of the phase of light, it does not require a triangulation base and can therefore be used to inspect, e.g., the bottom of a drill hole or functional surfaces that are surrounded by steep walls [23].

Figure 2 shows the basic setup of a digital holographic microscope for shape measurement of technical objects [24]. It consists of an illumination, a microscope objective, and optics to guide the reference wave. In our systems, we use optical fibers to transport illumination and reference waves in order to achieve maximum flexibility. If refocusing is necessary, it is useful to employ a telecentric microscope objective in order to avoid depth-dependent magnification [25].

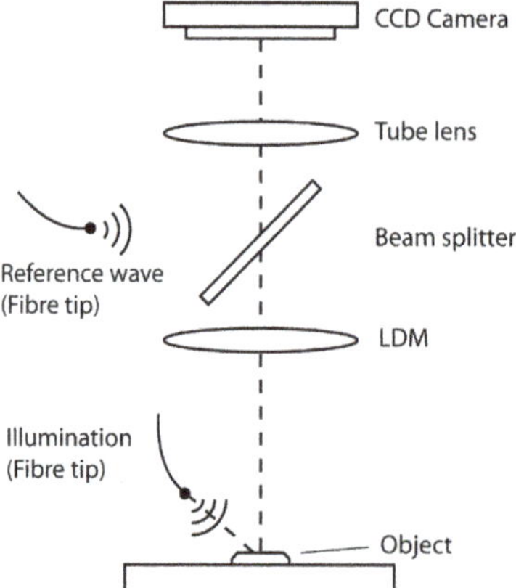

Figure 2. Digital holographic microscope setup: The reference wave and the illumination are provided by an optical fiber. The object and reference wave are superposed using a beam splitter between the long distance microscope objective (LDM) and the tube lens. From the recorded interference pattern, it is possible to refocus the object subsequent to the measurement by means of computational methods.

For microscopic investigations, it is common that the setup is arranged to image the object onto the camera target. Hence, the interference pattern across the camera plane allows determining the complex amplitude [11]

$$u(\vec{x}) = a(\vec{x}) \exp\left[i\phi(\vec{x})\right] \tag{1}$$

of the wave field across the object plane with its amplitude $a(\vec{x})$ and phase $\phi(\vec{x})$. The shape of the object $h(\vec{x})$ can be determined from the corresponding phase distribution $\phi(\vec{x})$, which depends on the optical path difference between the object and the reference wave. With a plane reference wave and normal illumination, we find

$$\phi(\vec{x}) = \frac{2\pi}{\lambda} \cdot 2h(\vec{x}), \tag{2}$$

with the wavelength λ. Because of the ambiguity of the phase, heights can only be uniquely determined within the interval $[-\pi, \pi]$. For configurations corresponding to Equation (2), this means a height range of $[-\lambda/4, \lambda/4]$. If this is too limited, multiple measurements ϕ_i associated with different wavelengths λ_i can be evaluated in combination [24]. The most prominent approach for this task is the so-called synthetic wavelength method [26], which is based on the phase difference of two phase distributions ϕ_1 and ϕ_2 corresponding to measurements with wavelengths λ_1 and λ_2

$$\phi_2(\vec{x}) - \phi_1(\vec{x}) = \frac{2\pi}{\Lambda} \cdot 2h(\vec{x}), \tag{3}$$

with the synthetic wavelength $\Lambda = (\lambda_1 \cdot \lambda_2)/(\lambda_1 - \lambda_2)$. This extends the interval of unique height values to $[-\Lambda/4, \Lambda/4]$, which can be significantly larger if the wavelengths are selected to be close together.

Finally, in microscopy, because of the large numerical aperture NA of the imaging system, the depth of focus $d_f = \lambda/(NA)^2$ is often smaller than the size of the object, leaving parts of it out of focus. In digital holography, we have access to the full complex amplitude shown in Equation (1) of the object wave and, therefore, can numerically scan through various object planes. For example, we can employ the plane wave decomposition to numerically propagate from plane z_0 to z_1 [27]

$$u(\vec{x}; z_1) = F^{-1}\{\hat{u}(\vec{x}; z_0) \exp\left(ik_z\Delta z\right)\}, \tag{4}$$

with $\Delta z = z_1 - z_0$, $\hat{u}$ being the Fourier transform of u, F^{-1} denoting the inverse Fourier transformation and k_z being the z-component of the k-vector in frequency space. In what follows, we will show two examples, where we used the above methods to arrive at a fast and robust in-line inspection using digital holographic microscopy.

Multiple Simultaneous Viewing Directions Using Four Digital Holographic Microscopes (MuVie)

The first example significantly extends the range of applicability of a measurement system based on the principle setup shown in Figure 2 with respect to the shape of the object under investigation. In order to achieve the goals of a fast measurement (<1 s/part), a measurement uncertainty < 5 µm, robustness against mechanical disturbances an and a 360° view of a complex geometry with a non-destructive, non-contact system, four key elements have been combined: (i) the method of digital holography, (ii) four directions of observation, (iii) two-wavelength contouring and (iv) the use of coherence gating.

The resulting setup of the Multiple-View holographic system (MuVie) is shown in Figure 3. It has been developed within the frame of project *B5* of the DFG funded collaborative research center *SFB 747 Micro Cold Forming* [22]. The system consists of four digital holographic units to provide sufficient observation directions to the same object to cover a 360° view. Each of the microscopes operates with two wavelengths $\lambda_1 = 636.55$ nm and $\lambda_2 = 642.10$ nm to yield a synthetic wavelength of $\Lambda = 73.64$ µm. The telecentric long distance microscope objectives have 10 times magnification and a numerical aperture of

$NA = 0.21$. The setup allows recording of multiple holograms at the same time through coherence gating. For this purpose, the system utilizes four laser diodes with a coherence length of approximately 1 mm. Using a system of optical fibers with appropriate lengths, interference is obtained between object and reference wave of the same holographic unit but impeded for light from different holographic units. The technique allows the holograms to be separated in Fourier space. For more details on this approach, see Reference [22].

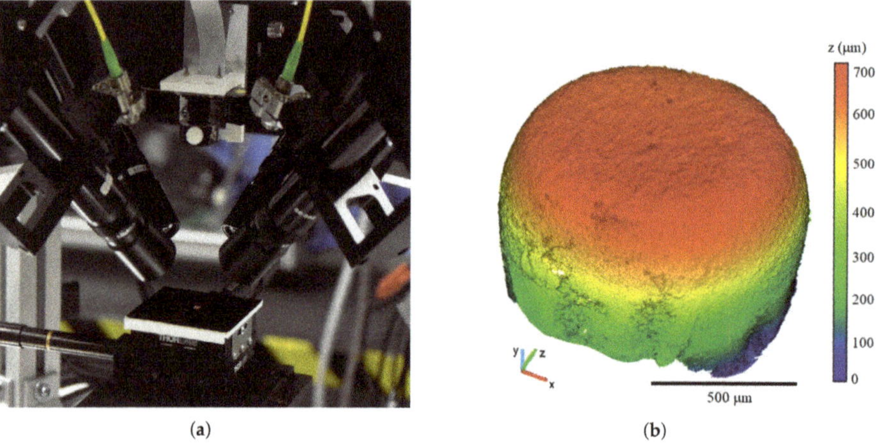

Figure 3. Digital holographic microscope for the inspection of micro cold formed micro cups: (**a**) The measurement system consists of 4 digital holographic microscopes to provide a full view of the top area of the object. (**b**) The measurement result shows a micro cup with diameter and height of approximately 1 mm. The lateral resolution is 2 µm over an extended range of 700 µm. The system can be used to quantify local imperfections, such as dents and deformations. The measurement uncertainty in the current configuration is 5 µm. Reprinted according to Creative Commons license from Reference [21].

The object is a micro cold formed micro cup made from aluminum with diameter and height of approximately 1 mm. The region of interest is the curved area around the top of the cup. As seen from the shape measurement in Figure 3b, the system can be used to quantify imperfections, such as dents, scratches and deformations. In the current configuration, it provides a lateral resolution of 2 µm and a measurement uncertainty of 5 µm. In the example shown here, the overall recording time is 50 ms. However, processing of the holograms requires approximately 15 s, due to refocusing and stitching requirements. To extend the depth of focus, we calculated the wave field across 10 object planes using Equation (4), with $\Delta z = 65$ µm.

Investigation of Functional Surfaces within the Cavity of a Micro Component

The second example is a digital holographic measurement system, which has been developed in collaboration with the company *VEW GmbH* to investigate functional surfaces within cavities. The system is shown in Figure 4a. It consists of a single microscope, which inspects the object in a normal observation direction at a rate of one part per second. Again two wavelengths $\lambda_1 = 632$ nm and $\lambda_2 = 637$ nm are used to arrive at a synthetic wavelengths of $\Lambda = 80.5$ µm. In contrast to the first system, the two holograms are recorded in quick succession using a fiber optical switch between the wavelengths. The recording time of a single hologram is approximately 1 ms to enable the system to also operate in a production environment. The object is a micro deep drawing part with a functional surface inside of a cylindrical tube, as shown in the inset of Figure 4a. The highlighted surface inside the cylinder is the region of interest. In the center of the object, we find a hole with a diameter of 0.9 mm. Figure 4b shows a measurement result with a clearly visible scratch in the upper left part of the surface. Similarly to the first system, the shape of the object can be determined with a measurement uncertainty of approximately 5 µm and a lateral resolution of 2 µm.

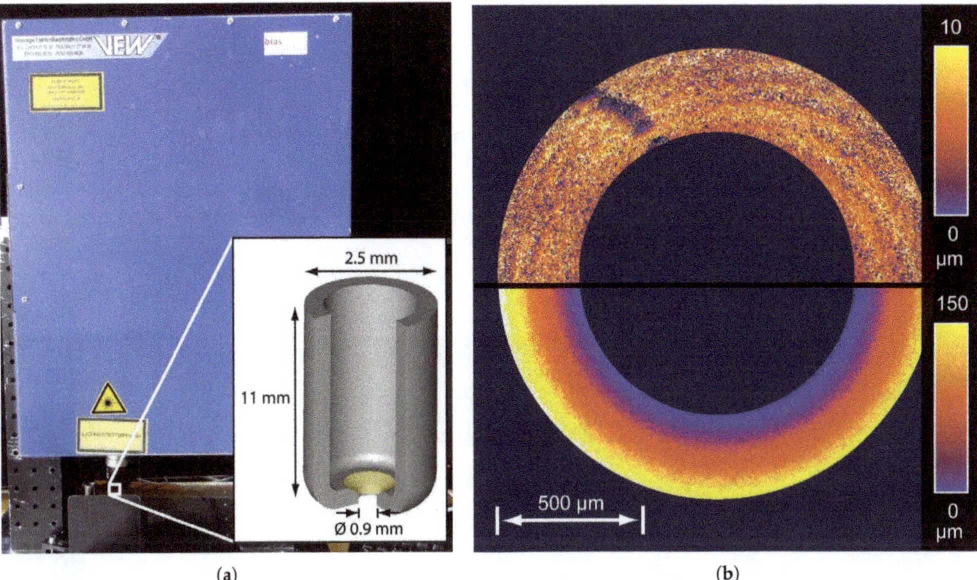

Figure 4. Digital holographic microscopy inside of cavities: (**a**) Digital holographic microscope, which is capable of testing one part per second. The inset shows a sketch of a typical object the system is designed for. The highlighted surface inside the cylinder is the region of interest. (**b**) A composed image of the same measurement result. The lower half depicts the measured profile, while the upper half shows the deviation from the known shape. We can clearly see a groove with a depth of approximately 5 µm, which indicates that this part is damaged.

3.1.2. Robust Interferometric Inspection Using Computational Shear Interferometry (COSI)

Shear interferometry has proven to be very robust against mechanical vibrations. Shear interferometers are imaging systems, which produce two identical images across the camera plane. The images are mutually shifted and the shift is referred to as the shear. In this configuration, no reference wave is required. However, the interference pattern formed by the two images provides information about finite differences (differential approach) across the surface of the specimen with interferometrical precision. Since the two interfering wave fields associated with the images travel almost identical optical paths, the system is intrinsically immune towards vibrations. An additional benefit of shear interferometry arises from its low demands with respect to both the temporal and the spatial coherence of light. This enables the use of eye-safe and cheap light sources, such as light emitting diodes or even arc lamps. In an industrial environment, where safety issues are critical, this is a decisive argument for shear interferometers. Finally, shear interferometers can also investigate objects with steep slopes, which would cause high fringe densities in common interferometric setups, and—as will be shown in the next section—they can operate with multiple light sources at the same time, which solves the major problem of a limited acceptance angle in interferometry.

One of the methods we use is Computational Shear Interferometry (CoSI) [28], which seeks to determine the underlying wave field associated with the two images. The benefit of CoSI is that it provides a wave field as a result. This enables almost any interferometric investigation, such as phase shifting interferometry, digital holography and quantitative phase contrast imaging, but with the advantage of using a robust shear interferometer [29].

Figure 5a shows the setup based on a liquid crystal spatial light modulator (SLM) [30]. The imaging system consists of a $4f$-arrangement with the SLM in the Fourier domain. It images any wave field incident in the input plane to the sensor plane where the camera is located. The SLM is birefringent, so that only light oriented along the slow axis will be modulated, whereas any other light will be reflected from the silicon back panel of the device [31]. We select the incoming light to be polarized exactly between the slow and the fast axis and program the SLM to display a blazed grating, which exhibits a single diffraction order. In this situation, two laterally separated images will appear across the sensor plane, where the separation, i.e., the shear, can be selected by the orientation and the magnitude of the blazed grating. The polarizers ensure that the two wave fields are able to form an interference pattern. The setup enables fast, precise and highly reproducible adjustment of the shear and ensures almost identical optical paths for both images.

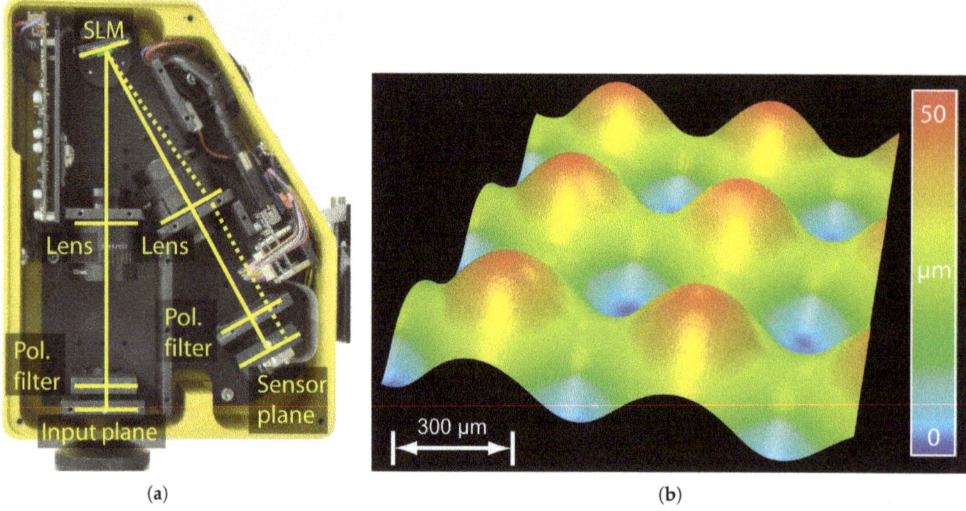

Figure 5. Computational Shear Interferometry: (**a**) Commercially available shear interferometer *Golden Eye*, which creates the shear using a liquid crystal SLM in the Fourier plane of a $4f$-configuration. The interferometer can be attached to any imaging optics and measures with interferometric precision while being insensitive against mechanical vibrations. (**b**) Measured shape of a micro lens array using CoSI in transmission. The system measures relatively steep slopes with a measurement uncertainty of $\sigma = 2.5$ nm. For more details, please refer to the text.

If we denote the wave field in the input plane as above by u and the shear by $\vec{s}$, we find that the intensity of the interference pattern in the camera domain is given by

$$I(\vec{x}) = |u(\vec{x})|^2 + |u(\vec{x}+\vec{s})|^2 + \mathcal{R}\{u^*(\vec{x})u(\vec{x}+\vec{s})\}, \tag{5}$$

where $\mathcal{R}$ means the real part. Using phase shifting techniques [32], we can isolate the argument of the real part and arrive at the cross amplitude $M(\vec{x}) = u^*(\vec{x})u(\vec{x}+\vec{s})$. We use typically $n = 5$ to 10 cross amplitudes M_n associated with different shears $\vec{s}_n$, to minimize the functional [29]

$$L(f) = \sum_n ||M_n(\vec{x}) - f^*(\vec{x})f(\vec{x}+\vec{s}_n)||^2 \tag{6}$$

using an iterative steepest descent algorithm. The complex amplitude found is the minimum least squares solution for the wave field

$$u(\vec{x}) = \arg\min_f L(f). \tag{7}$$

in the sense of Equation (6). With the complex amplitude $u(\vec{x})$, we can employ, for example, Equation (2), to determine the profile of the specimen. Using multiple measurements with varying wavelengths, we can also define a synthetic wavelength Λ and use Equation (3) to overcome limitations rooted in the ambiguity of the phase.

As an example, Figure 5b shows the measured shape of a micro lens array. The lenses have a diameter of 537 µm and relatively steep slopes. We have measured the array in transmission, using a $10\times$ microscope objective with a numerical aperture of $NA = 0.21$. The light source was a fiber coupled LED with a wavelength of 625 nm and a fiber diameter of 200 µm. The measurement uncertainty (1σ) was $\sigma = 2.5$ nm, even though the setup is immune against vibrations.

3.1.3. Multiple Aperture Shear Interferometry (MArS)

A very promising technique based on shear interferometry is *Multiple Aperture Shear Interferometry (MArS)*, which is currently under development in collaboration with the *Physikalisch-Technische Bundesanstalt* (PTB). The aim of MArS is to compensate for the problem of a limited acceptance angle of current interferometers for lens testing. If lenses with steep slopes are investigated, only a small portion of the light reflected by the surface under test (SUT) is incident into the entrance pupil of the interferometer. Usually, computer generated holograms (CGH) are used to pre-shape the illuminating wave front, so that the reflected light can be imaged [33]. However, CGHs are expensive and not customizable for different lenses.

MArS exploits the fact that shear interferometers can be operated with multiple independent light sources, i.e., multiple illumination apertures [34]. In this situation, a large number of light sources can be distributed around the SUT, and each light source will partially contribute to the measurement. Figure 6a shows a sketch of the setup. It consists of a shear interferometer, similar to the arrangement introduced in Figure 5a, but instead of a single light source, we use multiple illumination apertures, where two of them are evenly transported through the observation lens. Figure 6b depicts the intensity in the image plane for the case of a spherical mirror such as SUT [35]. Each light source creates a rectangular spot across the image of the SUT. Together, they cover the entire surface.

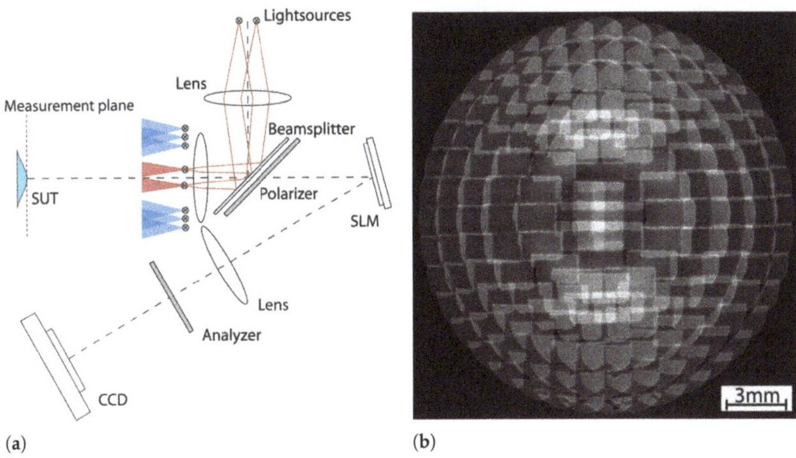

Figure 6. Multiple Aperture Shear Interferometry (MArS): (**a**) The setup consists of a shear interferometer and multiple independent light sources for illumination (illumination apertures). Some of the illumination apertures have to be guided through the observation lens in order to access the entire surface under test (SUT). (**b**) Intensity image recorded by the camera with a spherical mirror as the SUT. The illumination apertures appear as rectangular areas, where one aperture corresponds to one light source. Reprinted with permission from Reference [35].

When multiple independent wave fields are incident in the object plane, we cannot assign a consistent wave front to the light. However, we can still assign a coherence function

$$\Gamma(\vec{x}_1, \vec{x}_2) = \langle u^*(\vec{x}_1, t) u(\vec{x}_2, t) \rangle = \lim_{T \to \infty} \frac{1}{T} \int_{-T/2}^{T/2} u^*(\vec{x}_1, t) u(\vec{x}_2, t) dt \tag{8}$$

to it that depends on the two positions $\vec{x}_1$ and $\vec{x}_2$ in space. Shear interferometers measure the coherence function $\Gamma(\vec{x}, \vec{x} + \vec{s})$. In fact, for the fully coherent case, Equation (8) becomes the cross amplitude M introduced earlier. For the sum $u(\vec{x}, t) = \sum u_n(\vec{x}, t)$ of multiple independent wave fields $u_n(\vec{x}, t)$, we find after insertion into Equation (8)

$$\Gamma(\vec{x}, \vec{x} + \vec{s}) = \sum_n u^*(\vec{x}) u(\vec{x} + \vec{s}). \tag{9}$$

In MArS, we perform a large number of measurements while varying the shear $\vec{s}$ along a lattice so that we can interpret Equation (9) as a function in $\vec{s}$ instead of the spatial coordinate $\vec{x}$. If we select the shears across a small local area, we can regard Equation (9) as a sum of plane waves, with each of the waves being multiplied with a constant $u_n(\vec{x})$. The plane waves can be associated with wave vectors $\vec{k}_n(\vec{x})$, where each of the superposing wave fields contributes with one wave vector. We can separate the k-vectors in frequency space using the Fourier transform $F\{\Gamma(\vec{s})\}$ [36].

The idea is now to assign light rays to the directions given by the aggregate of wave vectors. The light rays can then serve as input to a variational problem, where an inverse ray-tracer is employed to determine the SUT [37], subject to the position of the light sources, which have to be known in advance by means of a calibration process. To solve the inverse problem, we use a method based on simulated annealing [38].

Figure 7 shows an example of a measurement [35]. The SUT in this case is an aspheric lens with a diameter of 25.4 mm and a design curvature radius of 49.8 mm. Figure 7a shows the aggregate of k-vectors, where different colors indicate that the light originated from different light sources. A total of 500 measurements with varying shears have been recorded to achieve this result. The object has been illuminated by 156 LEDs arranged around the SUT and 11 LEDs that have been projected through the lens. In Figure 7b, we see the measured shape of the SUT after 2000 iterations of the inverse ray-tracing approach. After subtraction of the best fit sphere, the deviation of the measured residual from the residual design topography was Peak-to-Valley (PV) 4 µm. These results are promising and show that the principle works. Currently, we are working on an improved calibration procedure in order to reduce the PV value by at least one order of magnitude, which is the requirement to introduce the system into the market of aspheric lens testing.

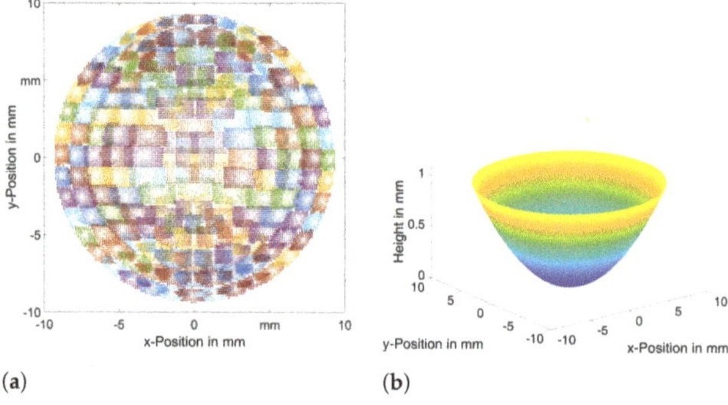

(a) (b)

Figure 7. Measurement results for an aspheric lens as the surface under test: (**a**) An aggregate of wave vectors extracted from the measured coherence function. Different colors indicate light from different light sources. (**b**) Calculated profile of the lens after 2000 iterations of the inverse ray-tracer. Reprinted with permission from Reference [35].

3.2. Techniques Based on Geometrical Optics Suitable for In-Process and In Situ Applications

Deflectometry, formerly often denoted as Fringe Reflection (FR) and Fringe Projection (FP), have proven their suitability for precise form measurement [14]. Deflectometry is suitable for reflective surfaces and has demonstrated its usefulness not only for the detection of surface and surface-near defects [15], but also for form measurement in optical manufacturing and qualification [39], achieves high precision [40] and—as a gradient-based method—is precise and robust without using a laser light source [20] that may introduce concerns with respect to eye safety. Although the measurement principle is simple, adjusting the setup and proper evaluation is involved in order to avoid systematic errors. Recent results for simpler and more effective calibration procedures applied to FP [41] established recently at our institute may also foster a breakthrough for in-line measurements using deflectometry.

Fringe Projection is suited for diffuse scattering surfaces and has demonstrated its ability for precise, robust, in-process measurement of complex geometries [42]. It is well suited for many technical surfaces and is intrinsically benign with respect to complex surfaces. Data capturing can be comparatively fast. Using a 120 Hz fringe projection in combination with a camera that only requires 8 ms for one picture one just needs 32 ms to capture a phase distribution. Although further measurements are needed in the commonly employed hierarchical phase shifting approach, the measurement accuracy only depends on the first measurement that uses the highest spatial frequency. In the following two sub-chapters, we present one example for measuring large objects and one for small objects.

3.2.1. Measurement of Wing Shapes Using Fringe Projection

The demand for a reduction of the fuel consumption and therefore CO_2-emission of aircrafts requires new technologies. One of them is to minimize turbulences on wings by manufacturing the surface with high quality. In the framework of the EU-Project *CleanSky*, we have developed a metrology technique that allows for precise form measurement during the wing production process [43].

Within the FP7 context of the European Union's CleanSky program, carried out as BLADE (Breakthrough Laminar Aircraft Demonstrator in Europe) under management of Airbus between 2011 and 2017, technology was being developed to build airplane wings for passenger planes that allow Natural Laminar Flow (NLF) under cruise conditions. "Natural" in this context means unsupported wings by extra flow sheets through micro-bores. The manufacturing of the surface requires tight tolerances on wing waviness and roughness. During manufacturing of NFL surfaces, an adequate quality control is therefore mandatory. At the start of BLADE, no measurement system existed that was capable of measuring large areas of wing surfaces with high lateral and depth resolution. For the purpose of the project, CleanSky's BLADE flight lab, an Airbus A 340-300, was used. It is the first aircraft in the world combining a transonic laminar wing profile with a standard aircraft internal primary structure.

Our institute, together with the VEW Corp. Bremen, supported the BLADE demonstrator with the development and application of an innovative metrology system to measure and evaluate profile, waviness, steps, gaps, 3D disturbances, and surface roughness of the wings from the beginning of their manufacturing up to the final assembly phase [43,44]. The measurement system is an FP-based system, custom built for the specific requirements of wing metrology. The nominal field of view is $1300 \times 800 \ mm^2$, making it perfectly suited for measurements in inter-rib spaces, the nose and on the upper cover of the BLADE wings with high accuracy. It can be adjusted to any orientation desired, and its position can be registered with laser tracker assistance. This makes it possible to locate unconnected surface segments correctly in 3D space, thus eliminating the need for large overlaps between surface sub-areas. To prevent problems of specular reflection of a glossy finish, the system can be equipped with three or more recording cameras. The depth resolution, in particular, the ability of the system to retrieve waviness, has been tested with a reference artifact that was designed in collaboration between Airbus and the National Physical Laboratory (NPL) of the UK and certified at the NPL. The artifact contains numerous features that are relevant to the BLADE metrology program. The measurement system is required to detect waves of different lengths and amplitudes. The smallest wave present on the test artifact has an amplitude of 100 μm, and the smallest steps are 50 μm high.

As an example, we report on measurements of an Invar (nickel-iron alloy) Outer Mold Line (OML) tool (size 9×2.1 m), which will serve as the exterior form of the experimental wing cover, and provide indications of the achievable 3D resolution in individual and combined measurements, as well as estimates on the match of the tooling surface against the CAD model [45]. In Figure 8, the measurement system is presented while measuring above the OML tool. In this case, a stereo-camera setup with an LED projector is used. For complete coverage of the tooling surface, over 50 individual measurements were necessary, generating a very large collection of data amounting to over 170 million points.

Figure 8. Fringe Projection system in the measurement position over the outer mold line (OML) surface. This part serves as the female tool of the upper cover part of the natural laminar flow (NLF) wing section on the starboard wing of the Airbus A340 BLADE flight test aircraft. The wing tool is fixed by a frame support beneath. Reprinted with permission from Reference [45].

Figure 9 combines the measured point cloud and the nominal CAD data. Blue and red marked areas are slightly over the defined specifications. The surface was brought into the tolerance band over its entire surface by tweaking the frame support. After modifications were made, all measured points showed the surface to be within the specifications.

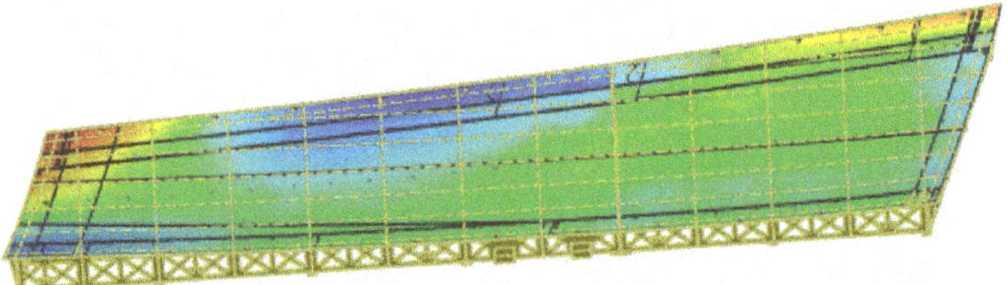

Figure 9. Comparison of measured mold tool surface sized 9×2.1 m with CAD data. The deviations are about 2 (red) and −2 mm (blue) caused by tweaking the frame support beneath the tool. Reprinted with permission from Reference [45].

Figure 10 presents views of the FP system on-site at Aernnova aircraft manufacturing facility in Berantevilla (Spain) in recording mode after the NFL wing final assembly phase (campaigns: GKN wing recording in 06-2016, Saab wing recording in 07-2016). The situations given in Figure 10 are exemplary for extreme measurement tasks for FP-based techniques. On the one hand, we have very complex geometries, and on the other hand, we have to deal with specular reflection of a glossy finish surface, which is not allowed to be manipulated. All measurements had to be carried out in-process while employees of the aircraft facility were working on the wings without interruption.

(a)

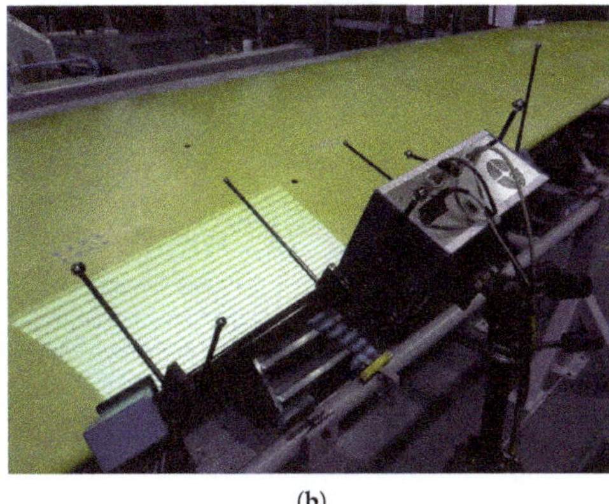

(b)

Figure 10. Fringe Projection measurements on wings. (**a**) Measurement of the BLADE GKN wing. Aerospace wing (front nose section and lower cover). Due to the size of the measuring field, the strongly curved front nose section, the subsequent geometry of the wing and recordings under daylight, we show here an example of an extreme measuring situation for the FP-technology. (**b**) Measurement of the BLADE Saab wing (front nose section and upper cover). The setup is equipped with retro-reflectors mounted on CFRP sticks for interacting with the laser tracker system. Reprinted with permission from Reference [44].

3.2.2. In-Process Measurement and Evaluation of Laser Added Manufacturing Processes for Metals

Additive manufacturing allowing layer-by-layer shaping of complex structures is of rapidly increasing interest in production technology. Especially for laser-based powder bed fusion processes of metals (PBF-LB/M), the authors of Reference [46] show a growing demand in production technology. As a consequence, there is a lively demand for quality inspection, process control and monitoring. There are more than 50 process parameters, adjustable or fixed concerning additive manufacturing (AM) technologies [47]. Among others, laser and scanning parameters, powder material and powder bed properties and recoating parameters are important for the PBF-LB/M. Many in-process sensing techniques such as cameras or pyrometers [48] for continuous temperature monitoring primarily focus on monitoring melt pool signatures. In Reference [49], a two-channel-pyrometer, coaxially integrated into the machine laser beam guidance, is implemented to identify thermal indicators with a resolution better than 10 μm. Other non-contact methods, for example, optical coherence tomography in selective laser melting [50] and selective laser sintering [51], are also applied. However, none of the methods mentioned have shown that the measured parameters have a direct relation to the internal quality or target geometry. A review of process defects and in situ monitoring methods in metal powder bed fusion may be found in [52].

As one of the important quality parameters, the presence of pores created during laser beam melting has to be avoided or minimized. However, currently there is no method for obtaining information on the pore density or its change during the process. As the density of the material decreases with an increasing fraction of pores, it should, however, be possible to detect a change of pore density by a precise height measurement of the structures. Figure 11 shows a schematic representation of the corresponding situation:

During the process, the volume of the initial powder shrinks due to the laser beam melting process. Therefore, the height of the recrystallized structure is lower than that of the powder bed surface. Due to the density dependence of the volume of recrystallized material on the volume of embedded pores, there is an additional slight height difference between structures that have a low and a high porosity. A measurement of this height difference should therefore give an indication for a change in the pore density of the current process.

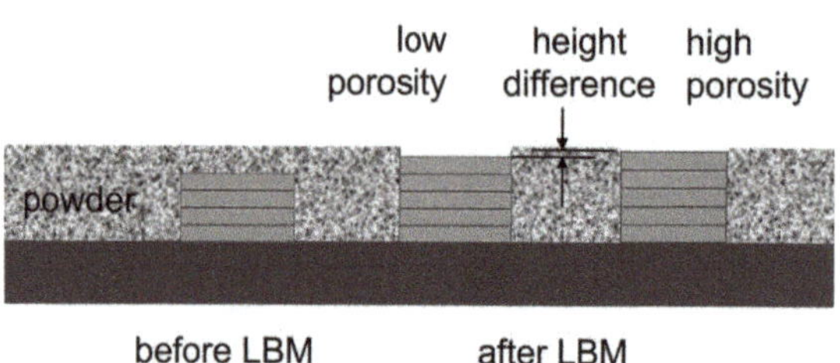

Figure 11. Schematic representation of the height change of structures due to laser beam melting. From left to right: An already existing structure is covered with metal powder ready for the next process step. The volume of the initial powder shrinks due to the processing. Therefore, the height of the recrystallized structure is lower than that of the surrounding powder bed surface. The shrinkage of the volume due to the melting process depends on the fraction of pores included in the treated material. Therefore, during growth of the structure, a small height difference evolves due to the density dependence of the volume of recrystallized material.

In Reference [53], we demonstrated a new approach to evaluate 3D laser printed parts in powder bed fusion-based AM in-line within a closed space. It was shown that high-precision metrology based on structured light is suitable for dimensional in-process sensing. The accuracy of the system presented was qualified to be below 10 μm in the z-coordinate perpendicular to the surface of the powder bed. This value is sufficient to reliably detect errors in powder coating or consolidation at a layer thickness of 50 μm or higher.

The basic idea of the measurement technology to be used in the PBF-LB/M process is the application of the Fringe Projection method. With this dimensional measurement technique, one can, for example, derive the shape, size and position as well as waviness and roughness. Figure 12 illustrates the method. The surface area, spread with metallic alloy powder and/or laser consolidated parts, is illuminated by structured light. Generally, a one-dimensional sinusoidal fringe pattern distribution is projected onto a surface and the modulated phase is calculated with one of the numerous techniques for analyzing fringes. The measuring principle is based on the phase shift method combined with triangulation [54]. The information of the surface profile is obtained from the deformed fringe pattern following the object shape. A system consisting of two identical CCD-cameras captures the fringe images projected onto the building plate inside the process envelope. The two (or more) cameras do not necessarily have to be housed in the printing chamber. With appropriate viewing windows, the system can also be installed outside. By marking the pattern, individual points on the surface can be recorded and calculated with precision in (x, y, z) coordinates using the camera's vision rays (sketched in blue in the picture). Our vision ray calibration technique for the cameras ensures the necessary high accuracy [41]. One advantage of the two-camera system implemented in this way is that the projector (outside the envelope) does not need to be calibrated.

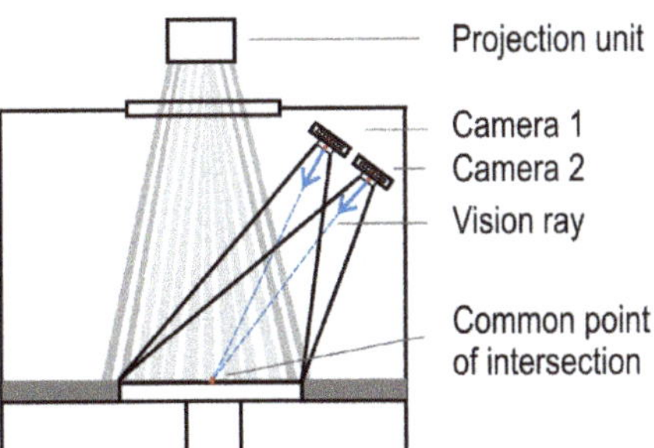

Figure 12. Principle of the fringe-projection technique used to analyze the PBF-LB/M process. A structured pattern is projected onto the powder bed by the projection unit. A stereo camera system (cameras 1 and 2) records this pattern. With the mathematical combination of the vision rays at a common point of intersection (sketched in blue as an example), an (x, y, z)-value of the surface can be obtained by triangulation for each pixel pair of the cameras. Reprinted with permission from Reference [53], where more information on the measurement principle can be found.

The following results give insights to the performance of our latest measurement setup and research methods (Obtained within the IGF-Project no. 21.178 N, for further details see section Funding). It demonstrates the state-of-the-art performance of dimensional metrology based on structured light applied to PBF-LB/M processes in-line. Figure 13a shows a CAD-design as a basis for the precise positioning of the cameras in the PBF-LB/M building chamber shown in Figure 13b. The arrangement in the CAD-design relative to the building plate corresponds to the exact scale as it has to be integrated into the PBF-LB/M System (SLM Realizer 250) at BIAS. In order to be able to achieve a very high lateral measurement resolution (in the in-plane × and y directions) of 35 µm, the measurement field should only depict the object to be evaluated. In addition, the fields of view of both cameras on the substrate must match as much as possible for the measurement principle to be implemented. The inner square within the gray ground plate in Figure 13a represents the size of a substrate of 150 × 150 mm. The measuring fields for the same system can, after a recalculation for the lenses, be expanded up to 280 × 280 mm on demand. The time required for the evaluation of a surface layer for this in-process measurement is about 5 s.

A special feature in the construction of the measuring system is the integration of the existing viewing windows of the PBF-LB/M system. The cameras are mounted and fixed in their final positions on a so-called "bridge" (compare Figure 13a). This camera bridge is installed under one of the two windows within the building envelope. The great advantage here is that the necessary cabling can be routed directly through the cover plate of the SLM chamber. The system closes off with the SLM chamber via a printed, heat-resistant hood (white part in Figure 13b with a height of only 90 mm). The fringe images are projected from outside the installation space through another viewing window of the PBF-LB/M system. That brings the further advantage that the projector does not have to be shielded from the process. No pressure loss in the building envelope after installation of the measuring system could be detected.

Figure 14a shows a recorded image of one camera from a series of measurements. It exactly shows what the camera observes coded in gray values without the fringe pattern. Cube shaped test consolidations were built with a size of 10 mm × 10 mm for each part with varying process parameters such as laser power and scanning speed, for details, see Table A1 of Appendix A. The scanning direction for all cubes is parallel to the y-axis of the SLM machine coordinate system (approximately vertical in Figure 14). A track is the result of the laser beam scanning along a straight line on the powder bed with a constant speed. The results presented below were taken from the layer 10 and 20 as stated in the following text and corresponding figures. In the complete printing procedure 40 layers have been formed.

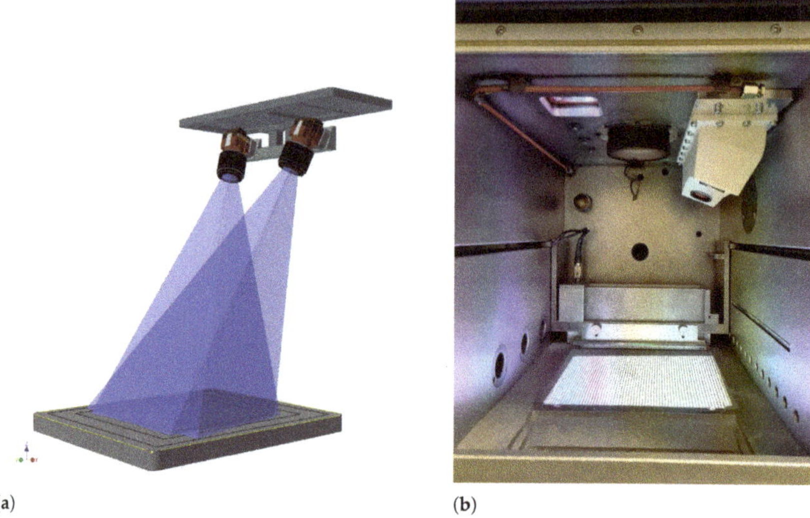

(a)   (b)

Figure 13. In-process measurement system. (**a**) CAD-Drawing of the dimensions and positions of the measuring system within the installation space. The calculated vision rays of the cameras on the building board are drawn in blue. For best practice, both cameras should capture the same field of view (here for the size of 150×150 mm). (**b**) Installation and test of the measuring system inside the Realizer 250 system of BIAS. A 3D printed, temperature-stable hood (white) shields the process. The reference pattern is projected from outside through a viewing window.

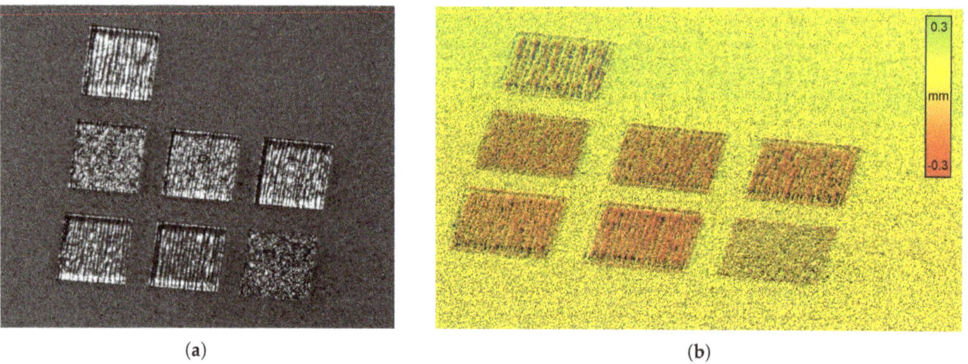

(a)   (b)

Figure 14. Results of printing depending on process parameters showing melted tracks of layer 10 with a size of 10×10 mm. (**a**) Perspective view of one measurement camera. (**b**) Visualized 3D data from the calculated point cloud. The measured geometry values show a high degree of similarity to the image of the camera. For height information, see the color scale. Process and sample parameters are given in Table A1 of Appendix A.

As the melting tracks shown in Figure 14a have a different appearance depending on the camera's perspective, one needs to calculate the point cloud of the final (x, y, z) coordinates of the structures created. These geometrical data are the product of the evaluation of both camera perspectives. The geometric coordinates of the point cloud are shown in Figure 14b in a view similar to that of the camera image in Figure 14a. It can be seen that the melting signatures of the consolidations are similar to the height profile.

A major goal of the research presented here is to derive conclusions about the quality of the printed parts from the measured dimensional values. In purely qualitative terms, it can be determined from the measurement data shown in Figure 14b that the tracks created by various process parameters have straight melted traces (e.g., upper and lower left), interruptions of the melted traces (e.g., middle row, center and right) or irregular patterns (middle row left or lower row right).

In order to determine the pore density as an important quantitative parameter of the process, we use X-ray computed tomography (X-CT). As described in Figure 11, one would expect a dependence of the structure height on the pore density. Figure 15 therefore shows the results of the porosity measured by X-CT as a function of the thickness difference between the structure with the highest porosity and the other structure heights. The height level of the structure with the highest porosity is therefore used as a reference level and set to zero. The results show a significant dependence of porosity on the thickness difference, as would be expected from the discussion of Figure 11. It should be noted, that the goal of this study is to determine the suitability of the measurement principle demonstrated here for an in-process characterization of the quality of the laser beam melting. An optimization of the PBF-LB/M process itself is beyond the scope of this study. Our results demonstrate, however, for the first time, the suitability of precise thickness evaluation using Fringe Projection for the in-process detection of the material quality created by an PBF-LB/M process.

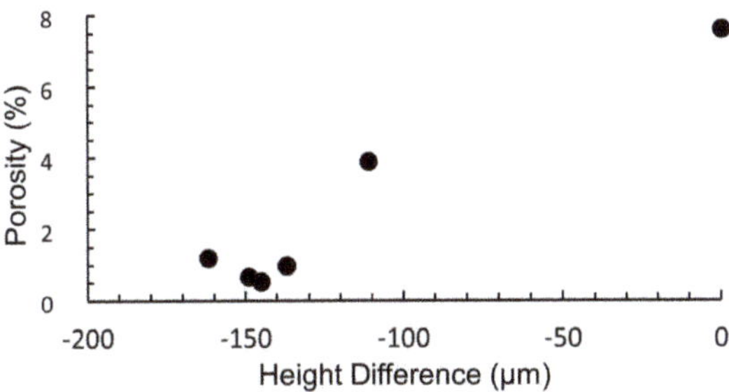

Figure 15. Porosity measured by X-CT of structures created by PBF-LB/M as a function of the height difference measured in-process by Fringe Projection. The values represent measurements using samples presented in Figure 14 after further growth to layer 20. The values of the height difference are taken from layer 20, the center of the cubes, which consist of 40 layers at the end of the growth process. For an easier overview, the material with the highest degree of porosity was used as a height reference with height difference = 0. Measured values follow the tendency that the height difference increases with the degree of sample porosity. The value of the structure located individually in the upper part of Figure 14a was not taken into account because the z-values do not appear plausible due to the large distance between the melted tracks (compare Figure 14b). Sample parameters are given in Table A1 of Appendix A.

For the assessment of the recorded data, usually surface parameters such as mean roughness (Sa), RMS roughness (Sq), Peak to Valley (PV) value (Sz), skewness (Ssk) or kurtosis (Sku) are determined according to Reference [55]. However, it turns out that the structure function (SF) [56] is much more useful for assessing 3D surface characteristics to reveal process deviations compared to the parameters stated above. The SF of order d depends on the separation parameter n for a one-dimensional discrete set of height values z_i with $i = 1, \ldots, N$ and is defined by

$$S_d(n) = \frac{1}{N-n} \sum_{i=1}^{N-n} (z_i - z_{i+n})^d. \tag{10}$$

The SF that represents the mean distance of each spatial distance (separation) does not refer to a statistical expectation of the data and is therefore independent of the measurement system. One of the important advantages that apply here is the significantly high sensitivity of the SF for roughness differences. In Reference [57], we therefore proposed a method for the in-process evaluation of powder beds used in PBF-LB/M based on the SF analysis. We have indicated that when comparing metallic powder height distributions, which differ only by 3 μm in their standard deviations, differences in roughness and waviness can be evaluated using SF. Our results [57,58] demonstrated that the SF can be interpreted as a criterion of the frequency content of the surface with respect to roughness and waviness, in which a periodicity within the sample surface is again represented by a periodicity in the SF. For more detailed definitions of the one- and two-dimensional SF, we refer to [59,60].

The following result demonstrates the successful assessment of the quality of the powder distribution on the basis of height values, which are shown in Figure 16 using the (x, y, z) values of layer 10, as described in the analysis above. The powder recoating procedure took place from right to left. For evaluation purposes, the geometric coordinates (z) of the point cloud are calculated perpendicular to the (x, y) plane. Two areas of 25 × 25 mm in size were selected out of the powder bed to compare and assess the even distribution. These areas are denoted as area 1 (right) and area 2 (left) in Figure 16. The calculated mean roughness Sa of the absolute of the ordinate values show a small difference of 1 μm only. The color values of the areas have been adjusted to make the height values more recognizable. Therefore, they have no relation to the color scale of Figure 16.

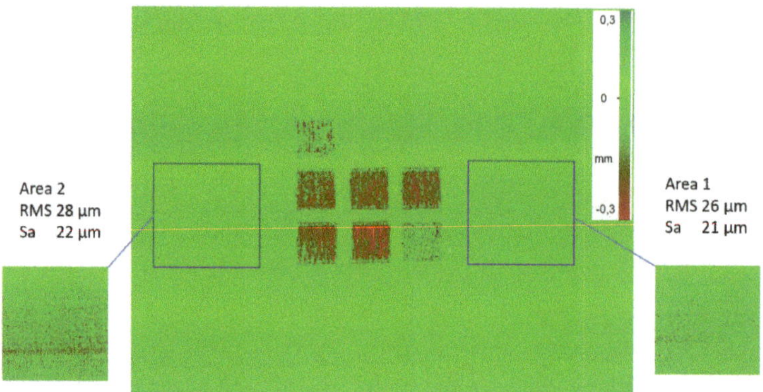

Figure 16. Presentation of the height distribution of layer 10 before powder recoating takes place from right to left. The measured sample has a size of about 150 × 150 mm (x, y). Two fields of the powder bed are chosen for investigation. For area 1 (**right**): Sa = 21 μm, RMS = 26 μm, powder appears rather smooth. For area 2 (**left**) : Sa = 22 μm, RMS = 28 μm, suggests slight ripples in the gray values. For each area, approximately 90,000 height values are obtained, and the lateral dimension is approximately 25 × 25 mm.

It is important to note that no powder features can be seen in the camera image (Figure 14). This means that a simple 2D photographic detection and evaluation does not provide any results here. In contrast, evaluation with height values is a good way of assessing the quality of the powder distribution.

Since the assessment with surface parameters such as Sa or RMS is only based on one value, the application of the structure function provides a wealth of information. In Figure 17, the calculated SF of both areas is documented. While the SF of area 1 runs relatively smoothly, the SF of area 2 shows the ripples of the powder distribution very clearly. In the high to mid spatial frequency range, both SFs show almost the same ordinate values. This confirms the similar roughness values presented in Figure 16. One can notice further that the two SFs differ a little from a separation value of around 150. This may be due to a slight slope in the powder distribution.

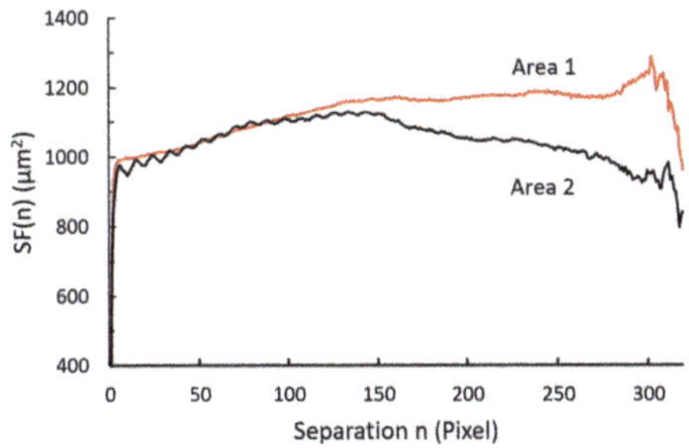

Figure 17. Multidirectional determined structure functions of area 1 (red) and area 2 (black). In the higher spatial frequency range up to about the separation n = 100, the amplitude of both SFs is almost identical. This clearly confirms the very similar roughness values. For further explanation, see text.

In summary, it can be stated that an evaluation of the PBF-LB/M AM technology using a non-contact measurement of (x, y, z) coordinates provides useful knowledge with regard to quality. The assessment of the consolidated material has shown that the mean height of a printed layer depends on its porosity. This result should now further be exploited and correlated with experiments using a wide range of parameters from other PBF-LB/M processes. As a second result, the uniformity of the powder application can be assessed down to the single-digit micrometer range. The statistical method of the structure function provides very precise and instructive results here.

4. Discussion

The following discussion describes some highlights on the conceptional progress to achieve fast, precise and rugged optical metrology for complex samples or measurement situations as discussed in the context of Figure 1. Speed (S) and Precision (P) have been continuously enhanced in the past until they meet physical or economic limits. In general, short illumination times and/or single shot measurements are not only beneficial for measurement speed but also for an increase in Robustness (R). We have concentrated here on *intrinsically* robust methods that lend themselves also to complex geometries (C). Gradient- or differential-based measuring principles are a clue to obtaining rugged measurement systems. Table 1 thus describes successful concepts discussed in this paper with respect to robustness and complexity.

Table 1. Overview over measurement methods discussed in this paper and their impact on robustness (R) and the ability to deal with complex measurement situations (C). Compare to the tetrahedron of optical metrology, as shown in Figure 1.

Requirement	Method	Principle Feature	Benefits
Robustness (R)	Computational Shear Interferometry (CoSi)	measures finite phase differences (defined by shear)	(i) enables interferometry using low coherent light, (ii) intrinsically robust against vibration
	Deflectometry	differential measurement	intrinsically robust against vibration
	Fringe Projection	triangulation	robust against ambient conditions
Complexity (C)	Multiple Viewing Directions for digital holography (MuVie)	multiple illumination and observation directions	enables 360°-view
	Computational Shear Interferometry (CoSI)	measures coherence function	(i) measures extended curvature due to variable shear, (ii) enables use of MArS
	Multiple Aperture Shear Interferometry (MArS)	simultaneous use of multiple light sources	(i) enables measurement of aspheres, (ii) solves aperture problem
	Fringe Projection	triangulation	suitable for complex sample geometry

For interferometry, vibrations are a major concern, since they disrupt phase measurements [7]. This problem is often targeted, e.g., by compensation measures [61] or fast data acquisition [62]. In contrast to such approaches, we are concentrating on intrinsically robust interferometry approaches, especially based on CoSI. Additional degrees of freedoms are gained from CoSI, as it allows for a better adaption to surface gradients by a proper choice of the shear. In terms of geometrical complexity, we have demonstrated approaches that not only concentrate on the geometrical complexity as such but also use the tailoring of the coherence of light to approach difficult geometric conditions as tackled by MuVie and MArS. A substantial further benefit results from MArS that solves the aperture problem by simultaneously using multiple light sources.

For geometrical methods, the situation appears a bit different. Deflectometry, being a differential measurement technique, is intrinsically robust against rigid body motion. It does not need vibration isolation and usually copes with ordinary in-door illumination. Due to its dependence on specular reflection, an aperture problem may, however, easily arise for free form surfaces. Fringe projection requires diffusely scattering surfaces and is thus less dependent on the curvature or aspect ratio or complexity of the test surface. Geometrically based optical metrology usually has a comparatively simple and cheap setup as compared to interferometric systems, is rugged, and use of incoherent light sources such as projectors or monitors lends itself to easy use in a fabrication environment. However, it requires substantial effort to turn geometrical metrology techniques into a precise and easy to use measurement tool. Here, our recent work on improved and faster calibration techniques for Deflectometry and Fringe Projection has shown significant progress in precision and speed of system calibration [41,63,64].

5. Conclusions and Outlook

Optical metrology for industrial applications needs to be precise, fast, robust and capable of coping with complex geometrical conditions. The paper emphasizes on progress in dealing with robustness and complexity as prime areas of improvement.

The development starting from traditional phase front measurements to the measurement of the full complex amplitude of a wave field comes to its limits. Tailoring the coherence properties of light according to sample requirements has been demonstrated in the Multi View Digital Holography (MuVie) microscope approach, which employs careful adjustment of the coherence properties of the light in order to separate holograms created simultaneously during the measurement. Already during the work using CoSI, but finally using Multiple Aperture Shear Interferometry (MArS), a new paradigm becomes mandatory, as the use of a multitude of independent light sources requires a description of light fields based on the coherence function.

Substantial progress is also demonstrated for measurement methods based on geometrical optics such as deflectometry and fringe projection. Due to faster and more accurate calibration, these techniques are now much better suited for metrology applications in an industrial environment. The examples shown in this paper demonstrate that optical metrology based on fringe projection can deliver reliable results even under such harsh conditions as inside a chamber for powder-based metal melting processes.

Further work quantifying the improvements of robustness and complexity is outside the scope of this paper as such a quantification depends on a large number of technical parameters. Further work is also needed in order to achieve faster measurement and data evaluation. Progress in Computer Vision [65], Compressed Sensing [66] and Machine Learning [67] are expected to lead to significantly improved performance.

Author Contributions: Conceptualization: R.B.B.; methodology: C.F., M.K.; software: C.F., M.K.; validation, C.F., M.K.; formal analysis: C.F., M.K.; investigation: C.F., M.K.; resources: R.B.B.; data curation: C.F., M.K.; writing—original draft preparation: R.B.B., C.F., M.K.; writing—review and editing: R.B.B.; visualization: R.B.B., C.F., M.K.; supervision: R.B.B.; project administration: R.B.B., C.F., M.K.; funding acquisition, R.B.B. All authors have read and agreed to the published version of the manuscript.

Funding: Research based on holography and shear interferometry has been funded by the projects *Quality Inspection and Logistic Quality Assurance of Micro Technical Manufacturing Processes* (B5) and *Inspection of Functional Surfaces on Micro Components in the Interior of Cavities* (T3) within the collaborative research center *Cold Micro Metal Forming* (SFB 747), the project *Optical surface metrology with spatially and temporally partially coherent light wave fields* (OPAL) project no. 258565427, and the project *Verfahren zur vollständigen Bestimmung kohärenter Wellenfelder nach dem Prinzip der Selbstreferenz* (WeSer), project no. 60045034, all funded by the Deutsche Forschungsgemeinschaft (DFG), and project *Robuste Digitale Holografie in der Produktionsumgebung* (RobuHoP), project no. ZF4587001LT8, funded by the Federal Ministry for Economic Affairs and Energy (BMWi) via the German Federation of Industrial Research Associations (AiF). Work concerning BLADE was carried out within the EU's CleanSky program under JTI-CS-2010-5-SFWA-03-004 and Grant Agreement 286745 and 641489 as well. We sincerely thank the EU for support of this research. We would like to thank Sebastien Dubois (EU Project Officer CSJU) and David Belfourd (Project Lead, Metrology and Assembly, Airbus UK) for collaboration and support within BLADE. The work concerning PBF-LB/M is funded in the framework of the IGF-Project No.: 21.178 N / DVS-No.: 13.3214 of the "Forschungsvereinigung Schweißen und verwandte Verfahren e.V." of the German Welding Society (DVS), Aachener Str. 172, 40223 Düsseldorf. Funded originates from the Federal Ministry for Economic Affairs and Energy (BMWi) via the German Federation of Industrial Research Associations (AiF) in accordance with the policy to support the Industrial Collective Research (IGF) on the orders of the German Bundestag.

Institutional Review Board Statement: Not applicable.

Informed Consent Statement: Not applicable.

Data Availability Statement: All data leading to the presented results are available on reasonable request from either the corresponding author R. B. Bergmann (bergmann@bias.de) or from C. Falldorf (falldorf@bias.de).

Acknowledgments: The authors gratefully acknowledge technical support from M. Agour, C. Kapitza and R. Klattenhoff, A. F. Müller and R. Tietjen and acknowledge the preparation of test samples by R. Dörfert, all at BIAS, and are thankful for the partnership with S. Wagner, A. Gesierich and W. Li from VEW Bremen in the BLADE project mentioned in the funding section and for the collaboration with industrial members of the committee associated with the IGF-project mentioned in the funding section.

Conflicts of Interest: The authors declare no conflict of interest. The funders had no role in the design of the study; in the collection, analyses, or interpretation of data; in the writing of the manuscript, or in the decision to publish the results.

Abbreviations

The following abbreviations are used in this manuscript:

AM	Additive Manufacturing
CoSi	Computational Shear Interferometry
DH	Digital Holography
FP	Fringe Projection
MArS	Multiple Aperture Shear Interferometry
MuVie	Multiple Viewing Direction Digital Holographic Microscopy
NA	Numerical Aperture
PBF-LB/M	Laser-based Powder Bed Fusion of Metals
SF	Structure Function
SLM	Spatial Light Modulator
SUT	Surface Under Test
X-CT	X-Ray Computed Tomography

Appendix A

Table A1. Parameters of processing (two top lines) and characterization (two bottom lines) of samples shown in Figure 14. For all samples, hatch distance = 70 μm, layer thickness = 50 μm. A hatch distance is the difference between adjacent tracks. The chosen parameters lead to 142 hatches per layer for one cube. The metallic powder used in the investigations was TruForm™ 718 (PraxAir), a high-strength, corrosion-resistant nickel-chromium material with a melting temperature of 1260–1336 °C. The values are taken from layer 20, the center of the cubes. For an easier overview, the level of the sample in the lower right (highest degree of porosity) was set to the value 0. The sample on the upper left of Figure 14a was not taken into account because the z-values do not appear plausible due to the large distance between the melted tracks (compare Figure 14b).

	Sample Location					
Row	**Middle**			**Lower**		
Position in Row	**Left**	**Center**	**Right**	**Left**	**Center**	**Right**
laser power (W)	110	140	170	170	200	80
scanning speed (mm/s)	500	500	500	400	400	500
mean height (μm)	−111	−145	−149	−137	−162	0
degree of porosity (%)	3.9	0.53	0.66	0.97	1.2	7.6

References

1. Gao, W.; Haitjema, H.; Fang, F.Z.; Leach, R.K.; Cheung, C.F.; Savio, E.; Linares, J.M. On-machine and in-process surface metrology for precision manufacturing. *CIRP Ann. Manuf. Technol.* **2019**, *68*, 843–866. [CrossRef]
2. Shimizu, Y.; Chen, L.C.; Kim, D.W.; Chen, X.; Li, X.; Matsukuma, H. An insight into optical metrology in manufacturing. *Meas. Sci. Technol.* **2021**, *32*, 042003. [CrossRef]
3. Vacharanukul, K.; Mekid, S. In-process dimensional inspection sensors. *Measurement* **2005**, *38*, 204–218. [CrossRef]
4. Takaya, Y. In-Process and On-Machine Measurement of Machining Accuracy for Process and Product Quality Management: A Review. *Int. J. Automat. Technol.* **2014**, *8*, 566–590. [CrossRef]
5. Bergmann, R.B. Computational Optical Metrology. In Proceedings of the International Conference on Optical and Photonic Engineering (icOPEN 2016), Chengdu, China, 26–30 September 2016; p. 13.
6. de Groot, P. Principles of interference microscopy for the measurement of surface topography. *Adv. Opt. Photon.* **2015**, *7*, 1–65. [CrossRef]
7. de Groot, P. Vibration in phase-shifting interferometry. *J. Opt. Soc. Am. A* **1995**, *12*, 354–365. [CrossRef]
8. Albertazzi, A. Interferometry in harsh environments. In *Optical Imaging and Metrology*; Osten, W., Reingand, N., Eds.; Wiley-VCH: Weinheim, Germany, 2012; pp. 369–391.
9. Häusler, G.; Leuchs, G. Physikalische Grenzen der optischen Formerfassung mit Licht. *Physikalische Blätter* **1997**, *53*, 417–422. [CrossRef]
10. Pavlíček, P.; Häusler, G. Methods for Optical Shape Measurement and their Measurement Uncertainty. *Optomechatronics* **2014**, *8*, 292–303. [CrossRef]
11. Schnars, U.; Falldorf, C.; Watson, J.; Jüptner, W. *Digital Holography and Wavefront Sensing*; Springer: Berlin/Heidelberg, Germany, 2016.
12. Fischer, A. Fisher information and Cramér-Rao bound for unknown systematic errors. *Measurement* **2018**, *113*, 131–136. [CrossRef]
13. Yoshizawa, T. *Handbook of Optical Metrology*; CRC Press: Boca Raton, FL, USA, 2009.
14. Bergmann, R.B.; Bothe, T.; Falldorf, C.; Huke, P.; Kalms, M.; von Kopylow, C. Optical Metrology and Optical Non-Destructive Testing from the Perspective of Object Characteristics. In Proceedings of the SPIE 7791, Interferometry XV: Applications, San Diego, CA, USA, 4–5 August 2010; p. 779102.
15. Bergmann, R.B.; Huke, P. Advanced methods for optical non-destructive testing. In *Optical Imaging and Metrology–Advanced Technologies*; Osten, W., Reingand, N., Eds.; Wiley-VCH: Weinheim, Germany, 2012; pp. 393–412.
16. Schuth, M.; Buerakov, W. (Eds.) *Handbuch Optische Messtechnik–Praktische Anwendungen für Entwicklung, Versuch, Fertigung und Qualitätssicherung*; Hanser: Munich, Germany, 2017.
17. Harding, K. (Ed.) *Handbook of Optical Dimensional Metrology*; CRC Press: Boca Raton, FL, USA, 2013.
18. Elandaloussi, F.; Mueller, B.; Osten, W. Determination of technological parameters in strip mining by time-of-flight and image processing. In Proceedings of the Optical Measurement Systems for Industrial Inspection, Munich, Germany, 16–17 June 1999; pp. 346–352. [CrossRef]
19. Sirohi, R.S. *Introduction to Optical Metrology*; CRC Press: Boca Raton, FL, USA, 2016.

20. Bergmann, R.B.; Burke, J.; Falldorf, C. Precision optical metrology without lasers. In Proceedings of the International Conference on Optical and Photonic Engineering (icOPEN 2015), Singapore, 14–16 April 2015; Volume 9524, p. 952403.

21. Agour, M.; von Freyberg, A.; Staar, B.; Falldorf, C.; Fischer, A.; Lütjen, M.; Freitag, M.; Goch, G.; Bergmann, R.B. Quality inspection and logistic quality assurance of micro technical manufacturing processes. In *Cold Micro Metal Forming*; Vollertsen, F., Kuhfuß, B., Thomy, C., Friedrich, S., Maaß, P., Zoch, H.W., Eds.; Springer Nature: Cham, Switzerland, 2020; pp. 256–274.

22. Agour, M.; Falldorf, C.; Bergmann, R. Spatial multiplexing and autofocus in holographic contouring for inspection of micro-parts. *Opt. Express* **2018**, *26*, 28576–28588. [CrossRef]

23. Simic, A.; Falldorf, C.; Bergmann, R. Internal Inspection of Micro Deep Drawing Parts Using Digital Holography. In Proceedings of the Imaging and Applied Optics 2016, Heidelberg, Germany, 25–28 July 2016; Optical Society of America: Washington, DC, USA, 2016; p. DW1H.3.

24. Falldorf, C.; Huferath-von Luepke, S.; von Kopylow, C.; Bergmann, R.B. Reduction of speckle noise in multiwavelength contouring. *App. Opt.* **2012**, *51*, 8211–8215. [CrossRef] [PubMed]

25. Agour, M.; Falldorf, C.; Staar, B.; von Freyberg, A.; Fischer, A.; Lütjen, M.; Bergmann, R.B. Fast Quality Inspection of Micro Cold Formed Parts using Telecentric Digital Holographic Microscopy. In Proceedings of the MATEC Web of Conferences, Auckland, New Zealand, 4–8 February 2018; EDP Sciences: Les Ulis, France, 2018; Volume 190, p. 15008.

26. De Groot, P.; Kishner, S. Synthetic wavelength stabilization for two-color laser-diode interferometry. *Appl. Opt.* **1991**, *30*, 4026–4033. [CrossRef]

27. Goodman, J.W. *Introduction to Fourier Optics*, 3rd ed.; Roberts and Company: Englewood, CO, USA, 2005.

28. Falldorf, C.; von Kopylow, C.; Bergmann, R.B. Wave field sensing by means of computational shear interferometry. *J. Opt. Soc. Am. A* **2013**, *30*, 1905–1912. [CrossRef] [PubMed]

29. Falldorf, C.; Agour, M.; Bergmann, R.B. Digital holography and quantitative phase contrast imaging using computational shear interferometry. *Opt. Eng.* **2015**, *54*, 024110. [CrossRef]

30. Falldorf, C.; Kopylow, C.; Juptner, W. Compact lateral shearing interferometer to determine continuous wave fronts. In Proceedings of the 2007 3DTV Conference, Kos, Greece, 7–9 May 2007; pp. 1–4.

31. Falldorf, C.; Osten, S.; Kopylow, C.V.; Jüptner, W. Shearing interferometer based on the birefringent properties of a spatial light modulator. *Opt. Lett.* **2009**, *34*, 2727–2729. [CrossRef]

32. Hariharan, P.; Oreb, B.F.; Eiju, T. Digital phase-shifting interferometry: A simple error-compensating phase calculation algorithm. *Appl. Opt.* **1987**, *26*, 2504–2506. [CrossRef]

33. Burge, J.H. Applications of computer-generated holograms for interferometric measurement of large aspheric optics. In Proceedings of the International Conference on Optical Fabrication and Testing, Tokyo, Japan, 2 August 1995; International Society for Optics and Photonics: Bellingham, WA, USA, 1995; Volume 2576, pp. 258–269.

34. Falldorf, C. Taking the next step: The advantage of spatial covariance in optical metrology. In Proceedings of the Imaging and Applied Optics 2016, Heidelberg, Germany, 25–28 July 2016; Optical Society of America: Washington, DC, USA, 2016; p. DW3E.1.

35. Müller, A.F.; Falldorf, C.; Lotzgeselle, M.; Ehret, G.; Bergmann, R.B. Multiple Aperture Shear-Interferometry (MArS): A solution to the aperture problem for the form measurement of aspheric surfaces. *Opt. Express* **2020**, *28*, 34677–34691. doi: 10.1364/OE.408979. [CrossRef]

36. Falldorf, C.; Hagemann, J.H.; Ehret, G.; Bergmann, R.B. Sparse light fields in coherent optical metrology. *Appl. Opt.* **2017**, *56*, F14–F19. [CrossRef]

37. Hagemann, J.H.; Falldorf, C.; Ehret, G.; Bergmann, R.B. Form determination of optical surfaces by measuring the spatial coherence function using shearing interferometry. *Opt. Express* **2018**, *26*, 27991–28001. [CrossRef]

38. Kirkpatrick, S.; Gelatt, C.D.; Vecchi, M.P. Optimization by simulated annealing. *Science* **1983**, *220*, 671–680. [CrossRef]

39. Burke, J.; Li, W.; Heimsath, A.; von Kopylow, C.; Bergmann, R.B. Qualifying parabolic mirrors with deflectometry. *J. Eur. Opt. Soc. Rap. Public* **2013**, *8*, 13014. [CrossRef]

40. Schachtschneider, R.; Fortmeier, I.; Stavridis, M.; Asfour, J.; Berger, G.; Bergmann, R.B.; Beutler, A.; Blümel, T.; Klawitter, H.; Kubo, K.; et al. Interlaboratory comparison measurements of aspheres. *Mess. Sci. Technol.* **2018**, *29*, 055010. [CrossRef]

41. Bartsch, J.; Sperling, Y.; Bergmann, R.B. Efficient vision ray calibration of multi-camera systems. *Opt. Express* **2021**, *29*, 17125–17139. [CrossRef] [PubMed]

42. Reh, T.; Li, W.; Gesierich, A.; Bergmann, R.B. Vision Ray Camera Calibration for Small Field of View. In Proceedings of the Deutsche Gesellschaft für Angewandte Optik (DGAO), Brunswick, Germany, 21–25 May 2013; p. A019-9. Available online: http://www.dgao-proceedings.de (accessed on 29 October 2021).

43. Kalms, M. WIMO (Outer Wing Metrology)—*Novel Strategies to Better Measure the Quality of Wing Surfaces—Final Report Summary*; European Commission: Brussels, Belgium, 2017. Available online: https://cordis.europa.eu/article/id/147160-novel-strategies-to-better-measure-the-quality-of-wing-surfaces (accessed on 29 October 2021).

44. Kalms, M. *WiMo (286745) & WIMCAM (641489) Outer Wing Metrology & Measurement Campaigns, Final Report Summary*; European Commission: Brussels, Belgium, 2017. Available online: https://cordis.europa.eu/docs/results/286/286745/final1-wimo-wimcam-publishable-summary.pdf (accessed on 29 October 2021).

45. Burke, J.; Gesierich, A.; Li, W.; Bergmann, R.B. Measurement of Mould Tool for Laminar-flow Carbon-fibre Composite Airplane Wing Cover. In *Photogrammetrie—Laserscanning—Optische 3D-Messtechnik: Beiträge der Oldenburger 3D-Tage*; Wichmann Verlag: Berlin, Germany, 2014; pp. 116–125.

46. ISO/ASTM. *Additive Manufacturing—Design—Part 1: Laser-Based Powder Bed Fusion of Metals*; ISO/STN; Iteh Standards: Newark, NJ, USA, 2019; Volume 52911-1:2019.

47. Spears, T.G.; Gold, S.A. In-process sensing in selective laser melting (SLM) additive manufacturing. *Integr. Mater. Manuf. Innov.* **2016**, *5*, 1–25. [CrossRef]

48. Doubenskaia, M.; Pavlov, M.; Grigoriev, S.; Tikhonova, E.; Smurov, I. Comprehensive optical monitoring of selective laser melting. *J. Laser Micro/Nanoeng.* **2012**, *7*, 236–243. [CrossRef]

49. Tyralla, D.; Seefeld, T. Thermal based process monitoring for laser powder bed fusion (LPBF). *Adv. Mater. Res.* **2021**, *1161*, 123–130. Available online: https://www.scientific.net/AMR.1161.123 (accessed on 29 October 2021). [CrossRef]

50. Neef, A.; Seyda, V.; Herzog, D.; Emmelmann, C.; Schoenleber, M.; Kogel-Hollacher, M. Low coherence interferometry in selective laser melting. *Phys. Procedia* **2014**, *56*, 82–89. [CrossRef]

51. Gardner, M.R.; Lewis, A.; Park, J.; McElroy, A.B.; Estrada, A.D.; Fish, S.; Beaman, J.J.; Milner, T.E. In situ process monitoring in selective laser sintering using optical coherence tomography. *Opt. Eng.* **2018**, *57*, 041407. [CrossRef]

52. Grasso, M.; Colosimo, B.M. Process defects and in situ monitoring methods in metal powder bed fusion: A review. *Meas. Sci. Technol.* **2017**, *28*, 044005. [CrossRef]

53. Kalms, M.; Narita, R.; Thomy, C.; Vollertsen, F.; Bergmann, R.B. New approach to evaluate 3D laser printed parts in powder bed fusion-based additive manufacturing in-line within closed space. *Addit. Manuf.* **2019**, *26*, 161–165. [CrossRef]

54. Geng, J. Structured-light 3D surface imaging: A tutorial. *Adv. Opt. Photonics* **2011**, *3*, 128–160. [CrossRef]

55. DIN EN ISO. *Geometrical Product Specification (GPS)—Surface Texture: Areal—Part 2: Terms, Definitions and Surface Texture Parameters*; DIN; Beuth-Verlag: Berlin, Germany, 2012; Volume 25178-2:2012-09.

56. Kolmogorov, A.N. The local structure of turbulence in incompressible viscous fluids at very large Reynolds numbers. *Dokl. Akad. Nauk SSSR* **1941**, *30*, 299–303. (In Russian)

57. Kalms, M.; Bergmann, R.B. Structure function analysis of powder beds in additive manufacturing by laser beam melting. *Addit. Manuf.* **2020**, *36*, 101396. [CrossRef]

58. Kalms, M.; Bergmann, R.B. In-line quality control using dimensional metrology of 3D metal parts printed by laser beam melting. In Proceedings of the Nondestructive Characterization and Monitoring of Advanced Materials, Aerospace, Civil Infrastructure and Transportation XIII, Denver, CO, USA, 3–7 March 2019; p. 109710N. [CrossRef]

59. Kreis, T.; Burke, J.; Bergmann, R.B. Surface characterization by structure function analysis,. *J. Eur. Opt. Soc. Rap. Public* **2014**, *9*, 14032. [CrossRef]

60. Kalms, M.; Kreis, T.; Bergmann, R.B. Characterization of technical surfaces by structure function analysis. In Proceedings of the Nondestructive Characterization and Monitoring of Advanced Materials, Aerospace, Civil Infrastructure, and Transportation XII, Denver, CO, USA, 4–8 March 2018; p. 1059924. [CrossRef]

61. Wiersma, J.T.; Wyant, J.C. Vibration insensitive extended range interference microscopy. *Appl. Opt.* **2013**, *52*, 5957–5961. [CrossRef]

62. Li, Y.; Kästner, M.; Reithmeier, E. Vibration-insensitive low coherence interferometer (LCI) for the measurement of technical surfaces. *Measurement* **2017**, *104*, 36-42. [CrossRef]

63. Bartsch, J.; Kalms, M.; Bergmann, R.B. Improving the calibration of phase measuring deflectometry by a polynomial representation of the display shape. *J. Europ. Opt. Soc.—Rap. Publ.* **2019**, *15*, 1–7. [CrossRef]

64. Bartsch, J.; Sperling, Y.; Bergmann, R.B. Qualification of holistic and generic camera-system calibration by fringe projection. In Proceedings of the Automated Visual Inspection and Machine Vision IV, Online, 20 June 2021; p. 117870G. [CrossRef]

65. Szeliski, R. (Ed.) *Computer Vision—Algorithms and Applications*; Springer: Berlin/Heidelberg, Germany, 2011.

66. Eldar, Y.; Kutyniok, G. (Eds.) *Compressed Sensing—Theory and Application*; Cambridge University Press: Cambridge, UK, 2012.

67. Bishop, C.H. (Ed.) *Pattern Recognition and Machine Learning*; Information Science and Statistics; Springer: Berlin/Heidelberg, Germany, 2006.

Article

In-Situ Evaluation of the Pitch of a Reflective-Type Scale Grating by Using a Mode-Locked Femtosecond Laser

Dong Wook Shin [1], Lue Quan [1], Yuki Shimizu [1,*], Hiraku Matsukuma [1], Yindi Cai [1,2], Eberhard Manske [3] and Wei Gao [1]

[1] Precision Nanometrology Laboratory, Department of Finemechanics, Tohoku University, Sendai 980-8579, Japan; shin.dong.wook.t6@dc.tohoku.ac.jp (D.W.S.); quan.lue.p3@dc.tohoku.ac.jp (L.Q.); hiraku.matsukuma.d3@tohoku.ac.jp (H.M.); caiyd@dlut.edu.cn (Y.C.); i.ko.c2@tohoku.ac.jp (W.G.)

[2] Key Laboratory for Micro/Nano Technology and System of Liaoning Province, Dalian University of Technology, Dalian 116024, China

[3] Department of Mechanical Engineering, Ilmenau University of Technology, 98693 Ilmenau, Germany; Eberhard.Manske@tu-ilmenau.de

* Correspondence: yuki.shimizu.d2@tohoku.ac.jp; Tel.: +81-22-795-6950

Abstract: Major modifications are made to the setup and signal processing of the method of in-situ measurement of the pitch of a diffraction grating based on the angles of diffraction of the diffracted optical frequency comb laser emanated from the grating. In the method, the improvement of the uncertainty of in-situ pitch measurement can be expected since every mode in the diffracted optical frequency comb laser can be utilized. Instead of employing a Fabry-Pérot etalon for the separation of the neighboring modes in the group of the diffracted laser beams, the weight-of-mass method is introduced in the method to detect the light wavelength in the Littrow configuration. An attempt is also made to reduce the influence of the non-uniform spectrum of the optical comb laser employed in the setup through normalization operation. In addition, an optical alignment technique with the employment of a retroreflector is introduced for the precise alignment of optical components in the setup. Furthermore, a mathematical model of the pitch measurement by the proposed method is established, and theoretical analysis on the uncertainty of pitch measurement is carried out based on the guide to the expression of uncertainty in measurement (GUM).

Keywords: diffraction grating; grating pitch; mode-locked femtosecond laser; laser diffraction; diffraction equation; measurement uncertainty analysis

Citation: Shin, D.W.; Quan, L.; Shimizu, Y.; Matsukuma, H.; Cai, Y.; Manske, E.; Gao, W. In-Situ Evaluation of the Pitch of a Reflective-Type Scale Grating by Using a Mode-Locked Femtosecond Laser. *Appl. Sci.* **2021**, *11*, 8028. https://doi.org/10.3390/app11178028

Academic Editor: Andreas Fischer

Received: 27 July 2021
Accepted: 28 August 2021
Published: 30 August 2021

Publisher's Note: MDPI stays neutral with regard to jurisdictional claims in published maps and institutional affiliations.

Copyright: © 2021 by the authors. Licensee MDPI, Basel, Switzerland. This article is an open access article distributed under the terms and conditions of the Creative Commons Attribution (CC BY) license (https://creativecommons.org/licenses/by/4.0/).

1. Introduction

A diffraction grating, which has periodic fine pattern structures on its surface, is one of the most important optical components often employed in many scientific and industrial fields. A diffraction grating can be employed as the scale for measurement in an optical encoder, in which the relative displacement between an optical head and a diffraction grating can be measured [1]. Since the fine pattern structures on a scale grating are employed as the graduations for measurement [2,3], the evaluation of the grating pitch is an important task to assure the performance of an optical encoder.

Many methods have been developed so far to evaluate the grating pitch of a scale grating. The observation of three-dimensional profiles of fine pattern structures on a scale grating by high-resolution measuring instruments such as critical-dimension (CD) scanning electron microscopes (SEMs) or atomic force microscopes (AFMs) is a direct and straightforward method [4–7]. However, this method is not suitable for the evaluation of the whole length of a scale grating due to the limited measurement throughput. On the other hand, a method utilizing the laser diffraction, in which the period of pattern structure can be evaluated by the angle of diffraction of a diffracted laser beams emanated from the grating surface under evaluation, is a promising one for the evaluation of the whole length of a scale grating [8–13]. Although the measurand capable of being evaluated in this method is limited to a mean of the grating period over the area where the measurement laser beam is irradiated (namely, this method is an indirect method for measurement of the pitch of a scale grating), it is suitable for the calibration of large area grating patterns due to its high measurement throughput and non-contact measurement apparatus. On the other hand, in the conventional methods based on the laser diffraction with a single-mode measurement laser beam, the information to be obtained in experiments are quite limited due to the nature of the laser diffraction; in most of the cases, a few-order diffracted beams can be obtained when projecting a measurement laser beam onto a scale grating. This limits the measurement accuracy of a grating pitch.

To address the aforementioned issues, a new concept for in-situ [14] pitch measurement employing an optical frequency comb laser has been proposed. An optical frequency comb laser contains a lot of modes equally spaced in the optical frequency domain [15]. Since each mode has a deterministic light wavelength, a group of the first-order diffracted beams having different angles of diffraction can be obtained. With the employment of a fiber detector and a spectrometer for the observation of the diffracted beams in the optical frequency domain, much more information can be obtained for the pitch evaluation based on laser diffraction. Stabile and accurate optical frequency of each optical mode in an optical frequency comb [16] is also expected to contribute to the accuracy improvement of the in-situ pitch measurement. The feasibility of the proposed concept has been verified through some experiments with a developed prototype setup [17]. However, there are some problems that need to be addressed for applying optical frequency comb as a light source for the laser diffraction method. Since an optical frequency comb has multiple frequency modes with a narrow spacing in the optical frequency domain, the neighboring diffracted beams overlap with each other, resulting in the difficulty of identifying each mode in a spectrometer [18]. In the previous study by the authors [17], a Fabry-Pérot etalon with a high free-spectral range was employed to expand the mode-spacing of the optical frequency comb so that each mode in the group of the first-order diffracted beams could be distinguished. However, this could reduce the information of the angles of diffraction of the diffracted beams, resulting in diminishing the benefit of the proposed concept. In addition, the misalignments of the measurement laser beam and a fiber detector composed of an objective lens and a single-mode fiber in the setup could be sources of uncertainty in measurement of the grating pitch. Theoretical investigation on the measurement uncertainty of the proposed concept is thus necessary while considering the influences of these optical misalignments, although it has remained a task to be addressed.

In this paper, following the previous study by the authors [17], a major modification is made to the prototype setup by removing a Fabry-Pérot etalon, while applying a new signal processing technique based on the weight of mass method to the detection of a peak wavelength for measurement of the grating pitch. A mathematical model of the pitch measurement is also established based on the modified setup by including some parameters related to optical misalignments. By using the established mathematical model, a measurement uncertainty analysis is carried out based on the guide to the expression of uncertainty in measurement (GUM). It should be noted that, regarding the concept of the in-situ pitch measurement, all the experiments described in this paper have been carried out in an ordinary laboratory room condition where the optical setup is not shielded from external disturbances such as illumination with fluorescent lights.

2. Methods for In-Situ Measurement of the Grating Pitch with a Diffracted Optical Frequency Comb Laser

Methods for measurement of the grating pitch with diffracted laser beams are based on the light diffraction equation that can be expressed as follows [19]:

$$P \sin \theta_{\text{in}} + P \sin \theta_{\text{diff}} = m\lambda \tag{1}$$

where P is the grating pitch, θ_{in} is the angle of incidence of the measurement laser beam, θ_{diff} is the angle of diffraction of the mth-order diffracted beam, and λ is the light wavelength of the measurement laser beam. It should be noted that θ_{in} and θ_{diff} are defined with respect to the normal of the grating surface. Figure 1a shows a schematic of the setup for measurement of the pitch deviation of a grating. The setup is mainly composed of a monochromatic laser source, a rotary table equipped with a high-precision rotary encoder, and a detector unit (not indicated in the figure). A scale grating can be mounted on the rotary table for pitch measurement. In the setup, a collimated laser beam from the laser source is projected onto the grating surface to generate diffracted beams. The configuration where θ_{in} becomes equal to θ_{diff}, as shown in Figure 1a, is referred to as the Littrow configuration [20]. Under the condition of the mth-order Littrow configuration with the Littrow angle θ_m, $\theta_{\text{in}} = \theta_{\text{diff}} = \theta_m$, and the following equation can be obtained from Equation (1):

$$P = \frac{m\lambda}{2\sin\theta_m} \tag{2}$$

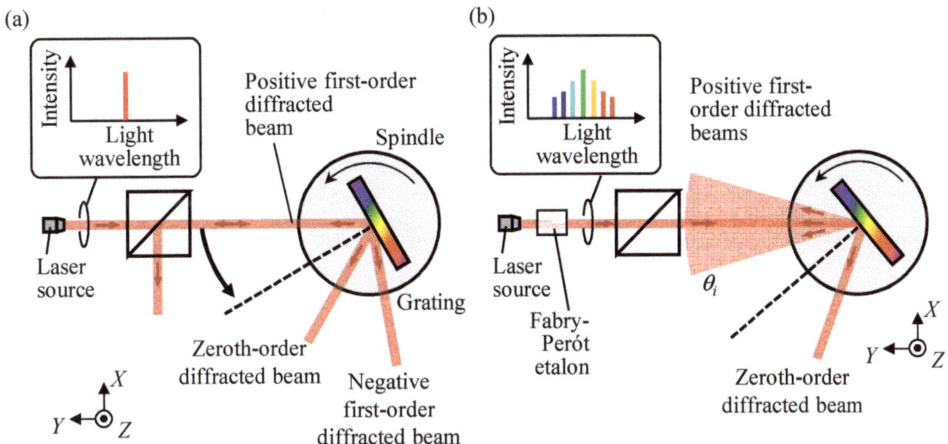

Figure 1. Laser diffraction method for measurement of a grating pitch: (**a**) Conventional method with a single-mode laser; (**b**) Proposed method with a mode-locked laser.

According to the above equation, the parameters need to be obtained for the calculation of P, θ_m and λ. Since λ can be treated as a known parameter, the grating pitch P can be evaluated by measuring θ_m in experiments. By rotating the grating under inspection with a rotary table equipped with a high-precision rotary encoder, multiple diffracted beams can be observed. For the detection of diffracted beams, image sensors such as a charge-coupled device (CCD) are often employed in the detector unit. In most cases, the number of diffracted beams available for pitch measurement is quite limited when employing a monochromatic laser source. For example, in the case where a scale grating having a nominal pitch of 1000 nm is evaluated with a laser beam having a light wavelength of 633 nm, the maximum m becomes 3; namely, only six diffracted beams (three positive diffracted beams and three negative diffracted beams) can be obtained. It should also be noted that the area to be irradiated by the measurement laser beam could vary with the increase of m. Namely, the grating pitch P evaluated by the diffracted beams with different diffraction orders could be from a different area on the scale grating.

These issues can be addressed by the method with an optical frequency comb laser [17]. Figure 1b shows a schematic of the optical setup for pitch measurement with an optical frequency comb laser. When the collimated laser beam of an optical frequency comb is projected onto a scale grating, the group of the positive first-order diffracted beams emanating from the projected area can be obtained [21]. It should be noted that the group of the negative first-order diffracted beams and those of the higher-order diffracted beams are not indicated in the figure for the sake of simplicity. The Littrow angle $\theta_{m,i}$ of the *i*th mode in the *m*th-order diffracted beam can be expressed by the following equation:

$$P_{m,i} = \frac{m}{2sin\theta_{(m,i)}} \times \frac{c}{v_i} = \frac{m}{2sin\theta_{(m,i)}} \times \frac{c}{i \times v_{rep} + v_{CEO}} \tag{3}$$

where v_i is the optical frequency of the ith mode, which can be represented by using the pulse repetition frequency v_{rep} and the carrier offset frequency v_{CEO} as $v_i = i \cdot v_{rep} + v_{CEO}$. Since a lot of Littrow angles can be obtained in the same diffraction order, the improvement of the measurement uncertainty with the averaging effect can be expected.

It should be noted that the diffracted laser beams should be observed in the optical frequency domain so that the light wavelength of each mode in the Littrow configuration can be determined. In the previous work by the authors [17], a Fabry-Pérot etalon with a high free-spectral range was employed so that each mode could be distinguished in the optical frequency domain. In this paper, a major modification is made to the optical setup; instead of separating the modes with the Fabry-Pérot etalon, a single-mode optical fiber was placed on the optical axis to specify the light wavelength mode corresponding to the Littrow configuration from the overlapped neighboring modes as shown in Figure 2. When a group of the diffracted beams was made incident to the optical fiber, the beams propagated through the core of the optical fiber by reflection on the boundary surface with the shelter part named clad. Due to the loss of the intensity from the reflection, the intensity spectrum consequently shows Gaussian distribution in which the light mode made parallelly incident to the optical axis shows the highest intensity. The spectrum of propagated beams within the mode field diameter (MFD), determined by the fall of the intensity to e^{-2}, was then detected by the optical spectrum analyzer, and the peak wavelength was determined by the weight of mass method. This modification enables the setup to fully utilize the information of the modes in the group of the diffracted beams.

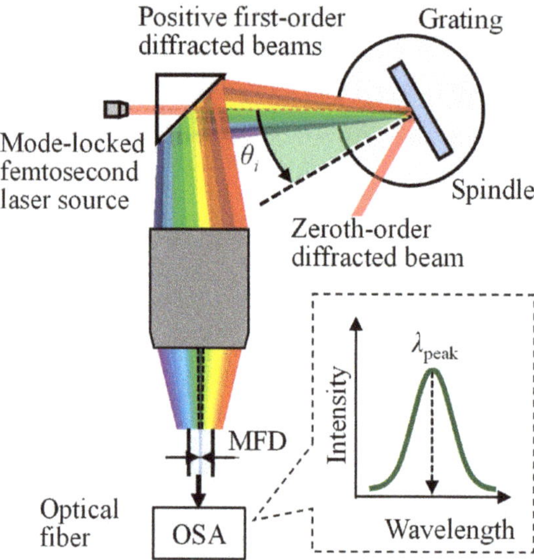

Figure 2. Peak wavelength detection for measurement of the grating pitch.

Figure 3a shows a three-dimensional drawing of the setup shown in Figure 1b. With the employment of a polarizing beam splitter (PBS) and a quarter-wave plate (QWP), the diffracted beams can be guided to the detector unit without going back to the laser source. The detector unit consists of an objective lens and a single-mode fiber, one side of which is connected to an optical spectrum analyzer (OSA). In the setup, the normal of the grating is aligned to be parallel with the X-axis when the angular position of the spindle is set to $\theta = 0°$. However, the measurement laser beam and the detector unit have angular misalignments with respect to the optical axis as shown in Figure 3b. Thus, variable α is given to the angular misalignment of the measurement laser beam about the Z-axis with respect to the X-axis, and variable β is given to that of the optical axis of the detector unit about the X-axis with respect to the Z-axis. Regarding these misalignments in the setup, the diffraction equation in Equation (2) for the first-order diffracted beams can be modified as follows:

$$P_{pos\_i} = \frac{\lambda}{\sin\theta_{\text{in}} + \sin\theta_{\text{diff}}} = \frac{\lambda}{\sin\left(\Delta\theta_{+1} - \frac{\alpha}{2}\right) + \sin\left(\Delta\theta_{+1} + \frac{\alpha}{2} + \beta\right)} = \frac{\lambda}{2\cos\left(\frac{\alpha-\beta}{2}\right)\sin\left(\Delta\theta_{+1} + \frac{\beta}{2}\right)} \tag{4}$$

$$P_{neg\_i} = \frac{\lambda}{\sin\theta_{\text{in}} + \sin\theta_{\text{diff}}} = \frac{\lambda}{\sin\left(\Delta\theta_{-1} + \frac{\alpha}{2}\right) + \sin\left(\Delta\theta_{-1} - \frac{\alpha}{2} - \beta\right)} = \frac{\lambda}{2\cos\left(\frac{\alpha-\beta}{2}\right)\sin\left(\Delta\theta_{-1} - \frac{\beta}{2}\right)} \tag{5}$$

where λ is the wavelength of the measurement beam, $\Delta\theta_{\pm1}$ is the angular position of the spindle, α is the angular misalignment of the measurement laser beam about the Z-axis with respect to the X-axis, and β is the angular misalignment of the optical axis of the detector unit about the X-axis with respect to the Z-axis.

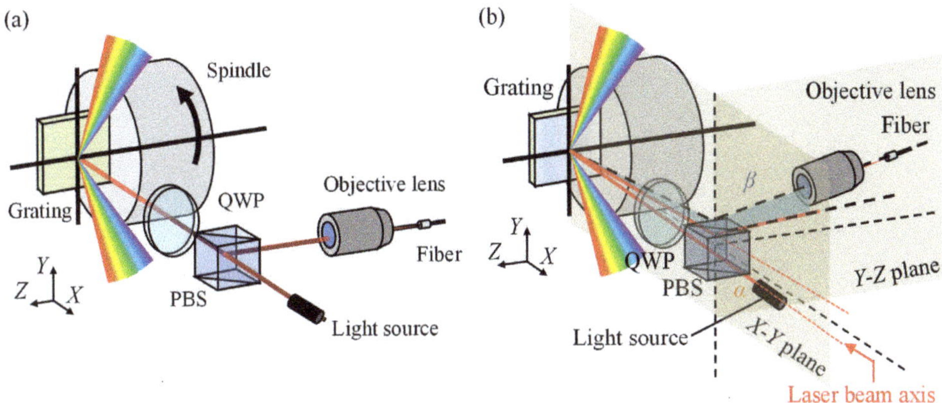

Figure 3. Optical configuration for measurement of the grating pitch; (**a**) Without the angular misalignments; (**b**) With the angular misalignments.

Based on Equations (4) and (5), the influences of α and β on measurement of the pitch deviation are estimated quantitatively. Figure 4a,b show the difference between the calculated pitch P_c obtained based on Equations (4) and (5) and the nominal pitch P due to the angular misalignments α and β, respectively. Calculations are carried out under the condition of $P = 833.33$ nm and an angular position of the spindle of 28°. The influence of α is estimated to be approximately -32 pm under the condition of $\alpha = 1°$, while that of β is estimated to be approximately 13.7 nm under the condition of $\beta = 1°$. From these results, it can be concluded that the corrections of α and β are necessary for accurate pitch measurement.

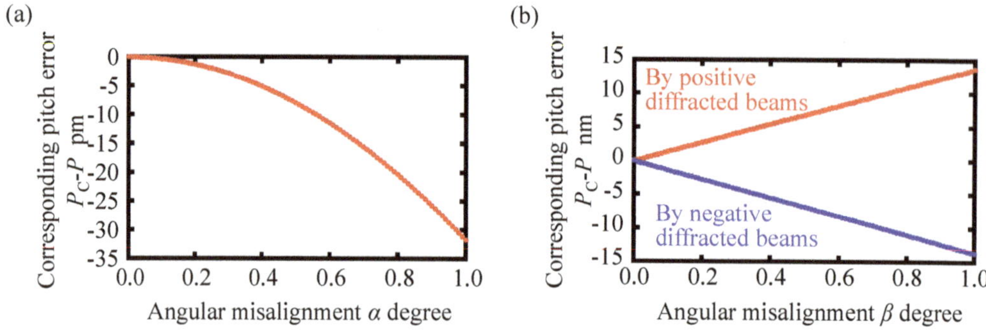

Figure 4. Influence of misalignments in the setup; (**a**) Influence of the angular misalignment of the measurement laser beam about the Z-axis (α); (**b**) Influence of the angular misalignment of the measurement laser beam about the X-axis (β).

3. Design and Development of the Optical Setup and Verification Experiment

A schematic of the setup employed in the following experiments is shown in Figure 5a. A fiber-based mode-locked femtosecond laser with a pulse repetition frequency ν_{rep} of 100 MHz was employed as the laser source. The laser beam from the laser source was guided into the setup by using an optical fiber. The laser beam from the edge of the fiber was collimated by a collimating lens, and was then projected onto a grating mounted on a spindle. It should be noted that a half-wave plate was placed just after the collimating lens in the optical path of the laser beam so that the polarization direction of the laser beam could be controlled. A pair of a quarter-wave plate and a polarizing beam splitter was employed as an optical isolator to prevent the diffracted beam from going back to the laser source. The diffracted laser beams, whose propagating directions were changed by the polarizing beam splitters, were captured by a detector unit composed of an objective lens and a single-mode fiber with a mode-field diameter of approximately 10.8 µm. One end of the single-mode fiber was placed at the focal plane of the objective lens, while the other was connected to a commercial optical spectrum analyzer (AQ6370D, Yokogawa Co., Ltd., Tokyo, Japan) so that the captured diffracted laser beams could be analyzed in the optical frequency domain. Figure 5b shows a photograph of the setup employed in the following experiments.

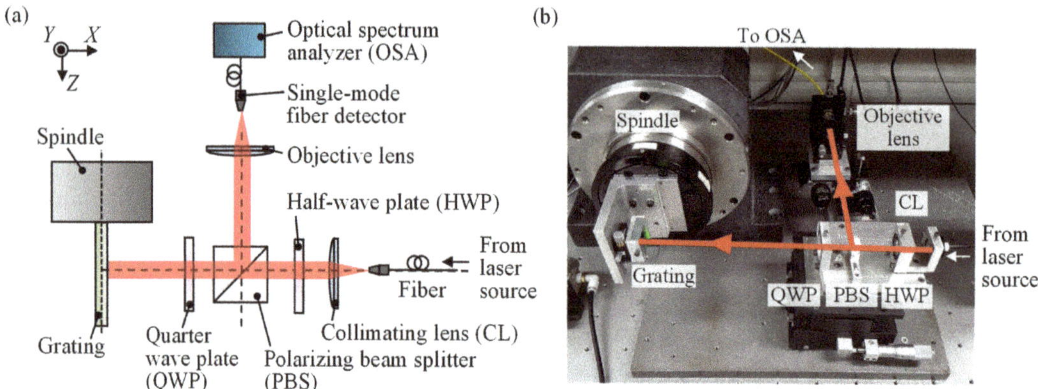

Figure 5. Experimental setup for measurement of the pitch of a reflective-type diffraction grating: (**a**) Schematic of the setup; (**b**) A photograph of the setup.

The detection of the angular position of the grating in the developed setup could strongly affect the pitch measurement by the proposed method. Therefore, at the beginning of the following experiments, the stability of the reading of the rotary encoder embedded in the spindle in the developed setup was evaluated. A variation of the rotary encoder reading was observed while the spindle was kept stationary at the positive and negative first-order Littrow configurations. Figure 6a,b show the variations of the rotary encoder reading at the angular positions $\theta = 28.08°$ and $-28.08°$, respectively, in a period of two hours. A sampling frequency of the encoder reading was set to be 10 Hz. Standard deviations of the rotary encoder reading were evaluated to be 0.06054 arc-second and 0.06475 arc-second for the positive and negative Littrow configurations, respectively. From these results, the spindle was confirmed to have positioning stability within the specification of the spindle (1 arc-second).

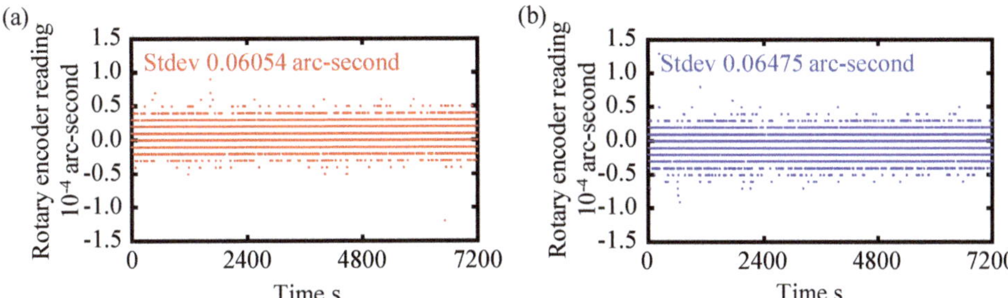

Figure 6. Reading of the rotary encoder embedded to the air-bearing spindle: (**a**) Positive first-order Littrow configuration; (**b**) Negative First-order Littrow configuration.

Alignments of the optical setup can be carried out by using the zeroth and first-order diffracted beams. Figure 7a shows a schematic of the setup. A retroreflector, which has the ability to reflect the laser beam to its source, is placed in front of the scale grating, while a beam profiler is placed at the focal plane of the objective lens. The reflected measurement laser beam is then focused on the detector plane at a point P_M in the beam profiler, and its Y-position is recorded. After that, the retroreflector is removed from the setup where the angular position of the spindle is set to $\theta = 0°$; this operation makes the zeroth-order diffracted beam from the scale grating be focused on the beam profiler at a point P_0. Meanwhile, with the existence of the angular misalignment α, the focused laser beam is made to shift along the Y-direction. According to the principle of laser autocollimation [22], the Y-directional spot displacement Δd on the beam profiler due to α can be expressed by the following equation:

$$\Delta d = F \tan \alpha \tag{6}$$

Since the focal length F of the objective lens was known to be 31 mm, α can be evaluated by detecting Δd with the beam profiler.

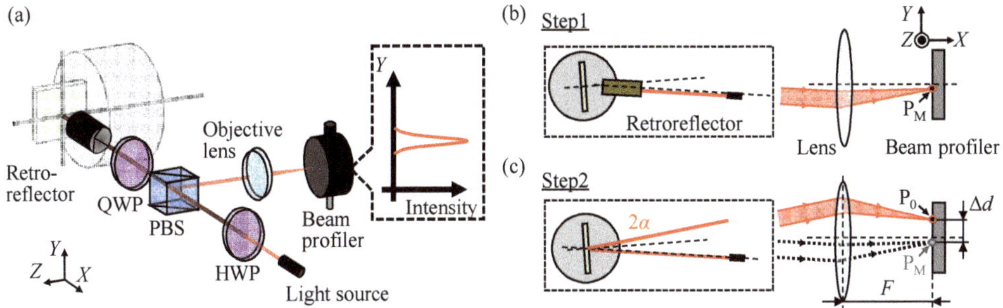

Figure 7. Alignment of the measurement laser beam by using a retroreflector and a beam profiler: (**a**) Setup with a retroreflector; (**b**,**c**) Procedure of the alignment.

Figure 8 shows the variations of the *Y*-positions of the measurement laser beam and the zeroth-order diffracted beam focused on the beam profiler in a period of 20 s. As the target grating, a holographic-type grating with a line density of 1200/mm was employed. From the mean values, Δd was evaluated to be 1.143 μm, and α was thus evaluated to be 7.579 arc-seconds based on Equation (6).

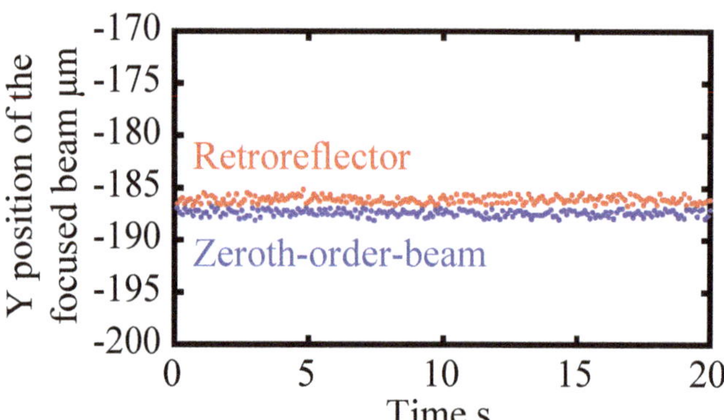

Figure 8. *Y*-position of the focused laser beam on the beam profiler during the alignment.

After the verification of the angular misalignment α, the alignment of the single-mode fiber in the fiber detector unit was carried out while the angular position of the spindle was kept stationary at $\theta = 0°$. Figure 9 shows the spectrum of the zeroth-order diffracted beam captured by the detector unit. The alignment of the single-mode fiber was carried out so that the total laser power to be detected by the detector unit became maximum while observing the spectrum of the captured laser beam. As can be seen in the figure, a non-uniform spectrum of the optical frequency comb laser was successfully observed.

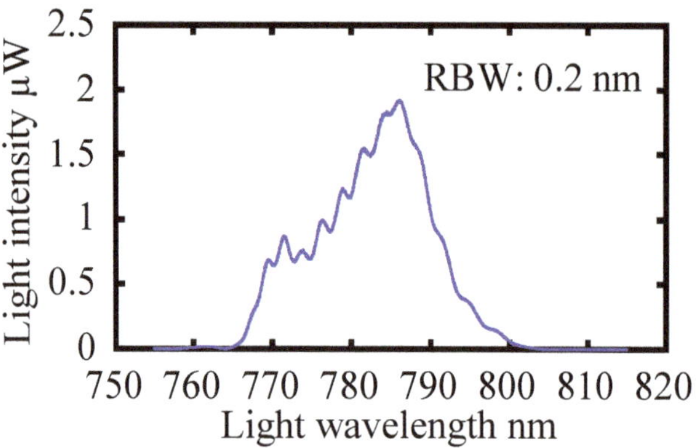

Figure 9. The optical spectrum of the mode-locked femtosecond laser employed in this paper.

Meanwhile, the detector unit had a small angular misalignment β about the *X*-axis even after the alignment; this can be evaluated in experiments by using the positive and negative first-order diffracted beams. Figure 10a,b show the variations of the spectra of the positive and negative first-order diffracted beams captured by the same detector unit when rotating the scale grating. In the figures, the spectra obtained at the angular position $|\theta|$ ranging from 27.48° to 27.52° in a step of 0.01° are plotted. As can be seen in the figure, the spectra of the negative first-order diffracted beams were found to shift toward the shorter light wavelength; this was mainly due to the influence of β, regarding Equations (4) and (5).

(a)

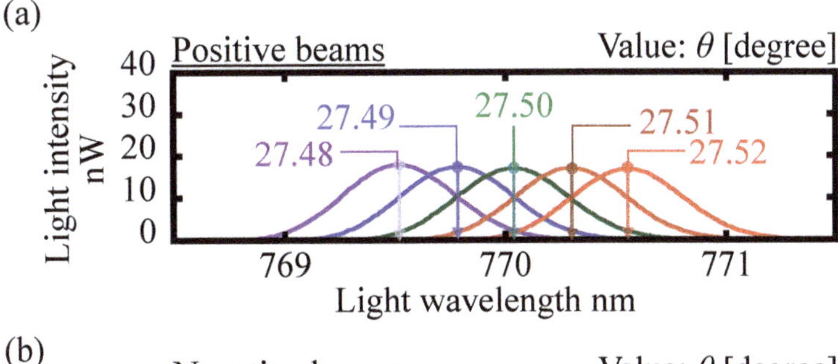

(b)

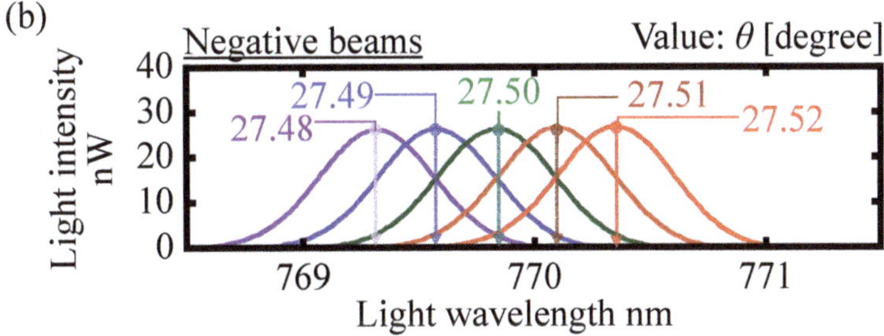

Figure 10. Spectra obtained at each angular position of the scale grating: (**a**) with the positive first-order diffracted beams; (**b**) with the negative first-order diffracted beams.

The grating pitch P can be evaluated by detecting the peak wavelength of the spectrum at each angular position of the spindle θ. In this paper, a weight-of-mass method based on the following equation was employed to detect the peak wavelength λ_{center} in a spectrum:

$$\lambda_{\text{Center}} = \frac{\sum\left(I_{\lambda_i} \times \lambda_i\right)}{\sum I_{\lambda_i}} \tag{7}$$

where λ_i is the ith sampled wavelength, and $I_{\lambda i}$ is the light intensity of the light wavelength component λ_i. Figure 11a shows the grating pitch calculated based on Equations (4) and (5) by the series of λ_{center} obtained at different angular positions θ of the spindle. It should be noted that the parameter β was set to be 0°. As can be seen in the figure, a discrepancy mainly due to the misalignment β was found between the grating pitches $P_{\text{pos\_}i}$ and $P_{\text{neg\_}i}$ obtained from the positive and negative first-order diffracted beams, respectively. To compensate for the influence of β, calculations based on Equations (4) and (5) were repeated by changing β in a step of 0.0001°. Figure 11b shows the sum of $|P_{\text{pos\_}i} - P_{\text{neg\_}i}|$ at each β. By performing fitting by sixth-order function approximation and finding the extremum, β that minimizes $\sum|P_{\text{pos\_}i} - P_{\text{neg\_}i}|$ was evaluated to be 0.001212° (4.363 arc-seconds). Figure 11c shows $P_{\text{pos\_}i}$ and $P_{\text{neg\_}i}$ calculated based on Equations (4) and (5) under the condition of $\beta = 4.363$ arc-seconds. As can be seen in the figure, a good agreement can be found between $P_{\text{pos\_}i}$ and $P_{\text{neg\_}i}$ after the compensation of the influence of β.

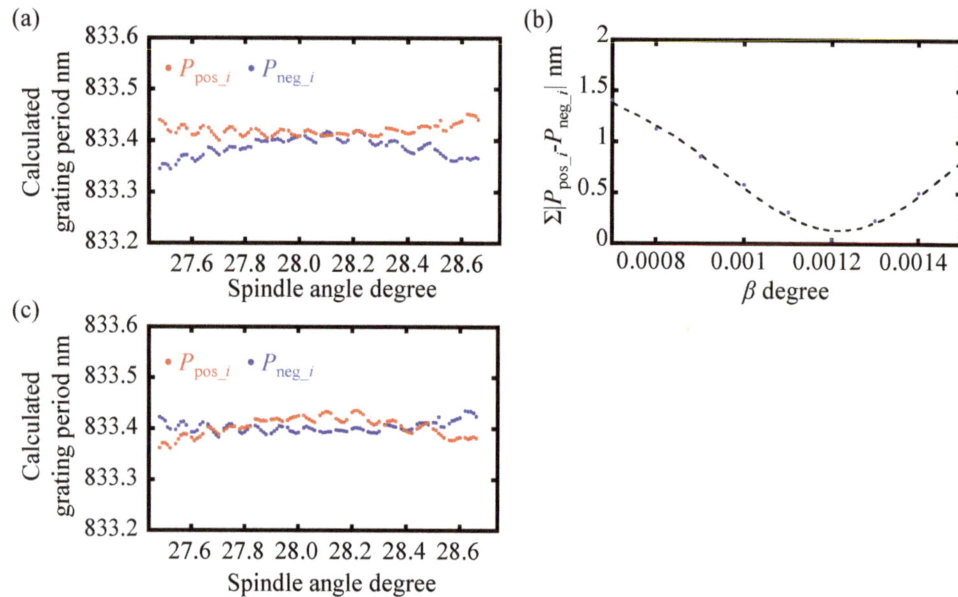

Figure 11. Grating pitch calculated from the peak wavelength obtained at each angular position of the scale grating: (a) Grating pitches P_{pos_i} and P_{neg_i} calculated by using the positive and negative first-order diffracted beams without the compensation of β ($\beta = 0$); (b) Variation of $\Sigma\,|\,P_{pos\_i} - P_{neg\_i}\,|$ with the change in β; (c) P_{pos_i} and P_{neg_i} after the compensation of β ($\beta = 4.363$ arc-seconds).

It should be noted that the optical frequency comb laser employed in this paper had a non-uniform spectrum, as can be seen in Figure 9. The non-uniform spectrum of the light source could affect the calculation of the peak wavelength based on the weight-of-mass method, resulting in the degradation of the uncertainty in pitch measurement. Compensation of the influence of the non-uniform spectrum of the mode-locked femtosecond laser beam employed in the setup was thus carried out by using the observed spectrum of the zeroth-order diffracted beam shown in Figure 9. In the compensation process, the spectrum of the first-order diffracted beam $I_{\pm 1}(\lambda_i)$ obtained at each angular position of the spindle was divided by the spectrum of the zeroth-order diffracted beam $I_0(\lambda_i)$ to obtain the normalized spectrum $I_{norm}(\lambda_i)$. Figure 12a,b shows the grating period P_{neg_i} calculated by using the spectra $I_{-1}(\lambda_i)$ and $I_{norm}(\lambda_i)$ of 100 optical modes observed in experiments, respectively. By the compensation process, the deviation of the calculated grating period P_{neg_i} obtained from the spectra in the wavelength range of 770 nm to 800 nm was reduced from 94.76 pm to 74.34 pm. This result demonstrated the effectiveness of the compensation process for the reduction of the uncertainty of pitch measurement. It should be noted that a slight difference can still be observed between P_{pos_i} and P_{neg_i}; The possible root causes contain the influence of the chromatic aberration of the optical components in the setup, as well as the instability of the spectrum of the mode-locked femtosecond laser beam, and further detailed investigation is required for the better understanding of the results.

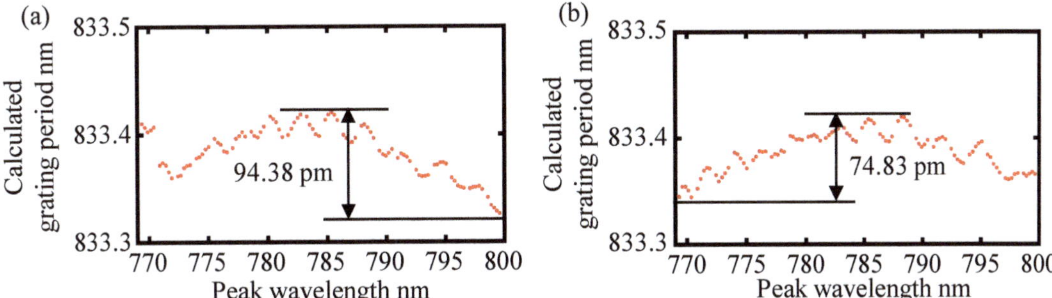

Figure 12. Compensation of the influence of the measurement laser beam spectrum. (**a**) Grating pitch calculated by using the spectra of the first-order diffracted beam $I_{-1}(\lambda_i)$; (**b**) Grating pitch calculated by using the normalized spectra $I_{norm}(\lambda_i)$.

4. Uncertainty Analysis

Measurement uncertainty analysis is also carried out based on the guide to the expression of uncertainty in measurement (GUM) [23] to verify the feasibility of the proposed method with the employment of an optical frequency comb laser. Since the mean of pitch values evaluated from multiple angles of diffraction of diffracted modes is evaluated in the proposed method, at first, the uncertainty of evaluating the pitch by a specific diffracted mode is evaluated. Then, the probability distribution of pitches from the angles of diffraction of entire modes was obtained to evaluate the measurement uncertainty of the proposed method.

Equations (4) and (5) can be modified as follows by considering the refractive index of air n_{air}:

$$P_{\text{pos\_}i} = \frac{\lambda_{\text{pos\_}i}}{2n_{air} \cos\left(\frac{\alpha - \beta}{2}\right) \sin\left(\theta_{\text{pos\_}i} + \frac{\beta}{2}\right)} \tag{8}$$

$$P_{\text{neg\_}i} = \frac{\lambda_{\text{neg\_}i}}{2n_{air} \cos\left(\frac{\alpha - \beta}{2}\right) \sin\left(\theta_{\text{neg\_}i} - \frac{\beta}{2}\right)} \tag{9}$$

In the following, the contribution of each parameter in the equations is evaluated.

(a) Angular misalignment α of the measurement laser beam about the Z-axis with respect to the X-axis ($u(\alpha)$)

At first, the contribution of α was evaluated. As described in the previous section of this paper, α was evaluated by using the zeroth-order diffracted beam reflected from the retroreflector; the Y-directional displacement of the focused laser beam on the CCD image sensor was converted into the angle α based on the principle of laser autocollimation [22]. Denoting the Y-positions of the focused laser beam on the image sensor when measuring the reflected beam from the retroreflector and the zeroth-order diffracted beam from the diffraction grating as Y_{retro} and Y_0, respectively, Equation (6) can be rewritten as follows:

$$\alpha = \arctan\left(\frac{Y_{retro} - Y_0}{F}\right) \tag{10}$$

Then, the standard uncertainty $u(\alpha)$, which is the contribution of α, can be evaluated by the following equation:

$$u_\alpha = \sqrt{\left[c_{Y\text{retro}} \cdot u_{Y\text{retro}}\right]^2 + \left[c_{Y0} \cdot u_{Y0}\right]^2 + \left[c_F \cdot u_F\right]^2} \tag{11}$$

where $u_{Y\mathrm{retro}}$, u_{Y0} and u_F are the standard uncertainties of Y_{retro}, Y_0 and F, respectively, while $c_{Y\mathrm{retro}}$, c_{Y0} and c_F are the sensitivity coefficients of Y_{retro}, Y_0 and F, respectively. Table 1 summarizes the contributions of each parameter on u_α. The Y-position of the focused retroreflected beam and zeroth-order beam on the profiler was obtained as a mean of 200 repetitive trials with a standard deviation of 0.333 µm and 0.299 µm, respectively. From Equation (10), the sensitivity coefficients $c_{Y\mathrm{retro}}$, c_0 and c_Y can be derived as follows:

$$c_{Y retro} = \frac{\partial \alpha}{\partial Y_{retro}} = \frac{\partial}{\partial Y_{retro}}\left[\arctan\left(\frac{Y_{retro}-Y_0}{F}\right)\right] = \frac{1}{F} \cdot \frac{1}{1+\left(\frac{Y_{retro}-Y_0}{F}\right)^2} \tag{12}$$

$$c_0 = \frac{\partial \alpha}{\partial Y_0} = \frac{\partial}{\partial Y_0}\left[\arctan\left(\frac{Y_{retro}-Y_0}{F}\right)\right] = \frac{-1}{F} \cdot \frac{1}{1+\left(\frac{Y_{retro}-Y_0}{F}\right)^2} \tag{13}$$

$$c_F = \frac{\partial \alpha}{\partial F} = \frac{\partial}{\partial F}\left[\arctan\left(\frac{Y_{retro}-Y_0}{F}\right)\right] = \frac{1}{1+\left(\frac{Y_{retro}-Y_0}{F}\right)^2} \cdot \frac{Y_0-Y_{retro}}{F^2} \tag{14}$$

Table 1. Uncertainty of the angular misalignment α.

Sources of Uncertainty	Symbol	Type	Uncertainty Value	Probability Distribution	Divisor	Standard Uncertainty u	Sensitivity Coefficient c	$\lvert c\rvert\cdot\lvert u\rvert$
Repeatability of the detection of Y_{retro}	u_{dretro}	A	0.333 µm	Gaussian	$\sqrt{200}$	0.0235 µm	0.116 arc-second/µm	2.731×10^{-3} arc-second
Repeatability of the detection of Y_0	u_{d0th}	A	0.299 µm	Gaussian	$\sqrt{200}$	0.0211 µm	-0.116 arc-second/µm	2.448×10^{-3} arc-second
Focal misalignment of the image sensor with respect to the objective lens	u_F	B	±0.5 mm	Rectangular	$\sqrt{3}$	0.289 mm	-4.193×10^{-3} arc-second/mm	1.211×10^{-3} arc-second
$u(\alpha)$								3.862×10^{-3} arc-second

By applying mean values of Y_{retro} (-186.160 µm), Y_0 (-187.303 µm) and a nominal value of F of 31.1 mm, $c_{Y\mathrm{retro}}$, c_0 and c_Y were evaluated as shown in Table 1. It should be noted that the misalignment of the image sensor with respect to the objective lens was considered to be within ±0.5 mm. By evaluating the contribution of each uncertainty source, $u(\alpha)$ was evaluated to be 3.862×10^{-3} arcsecond.

(b)   Angular misalignment β of the detector unit about the X-axis with respect to the Z-axis ($u(\beta)$)

The contribution of β was then estimated. As shown in Figure 11b, $\Sigma\,\lvert P_{\mathrm{pos}\_i} - P_{\mathrm{neg}\_i}\rvert$ corresponding to the change of β in a step of $0.0001°$ was calculated, then β which minimize $\Sigma\,\lvert P_{\mathrm{pos}\_i} - P_{\mathrm{neg}\_i}\rvert$ was selected by taking the extremum of the data curve obtained by sixth-order polynomial approximation fitting. However, as shown in Figure 11c, P_{pos_i} and P_{neg_i} did not perfectly match by this fitting process, and $\lvert P_{\mathrm{pos}\_i} - P_{\mathrm{neg}\_i}\rvert$ at each angular position of the spindle could contribute to the uncertainty $u(\beta)$ of evaluating β by polynomial approximation fitting. The deviation of β_i, which was obtained by the ith positive and negative Littrow angles was evaluated as shown in Figure 13. A standard deviation of $\lvert \beta\text{-}\beta_i\rvert$ was evaluated to be 1.875 arc-second. Since, 100 sets of P_{pos_i} and P_{neg_i} were employed, by applying a divisor of $\sqrt{100}$, $u(\beta)$ was evaluated to be 0.1875 arc-second.

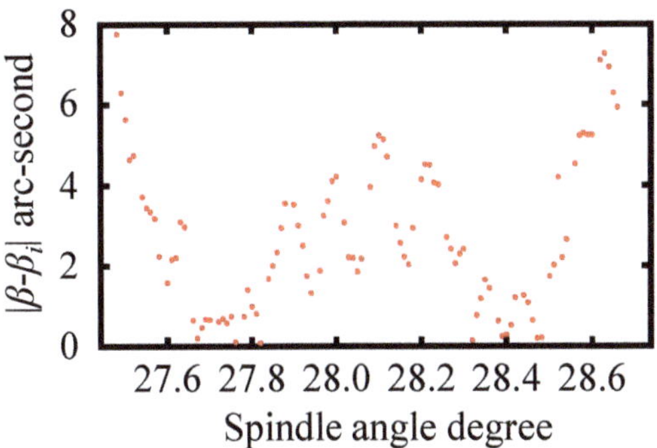

Figure 13. The deviation of β_i from β.

(c) Uncertainty of the detection of the angle θ_i of the *i*th-mode positive or negative first-order diffracted beam ($u(\theta_i)$)

In the proposed method, the *i*th-mode first-order Littrow angle θ_i was obtained from the readings of the optical rotary encoder embedded to the precision air-bearing spindle. Therefore, the resolution, stability and uncertainty of the reading of the optical rotary encoder could contribute to the uncertainty $u(\theta_i)$ of detecting the *i*th-mode first-order Littrow angle of the positive or negative first-order diffracted beam. The contribution of the resolution (u_{res}) was estimated from the information in the specification sheet of the rotary encoder provided by the encoder manufacturer. Meanwhile, the contribution of the stability of the encoder reading (u_{stb}) could somehow be affected by the stability of the developed setup, and was then evaluated by using experimental results. As can be seen in Figure 6a,b, standard deviations of the encoder reading when the scale grating was kept stationary at the angular positions of the positive and negative first-order Littrow configuration were evaluated to be 0.06054 arc-second and 0.06475 arc-second, respectively. Since a mean of 200 encoder readings at each angular position of the grating was employed as θ_i, by applying a divisor of $\sqrt{200}$ to the worse case (0.06475 arc-second), u_{stb} was evaluated to be 4.579×10^{-3} arc-second. Table 2 summarizes the contribution of each uncertainty source to $u(\theta_i)$.

Table 2. Uncertainty of the detection of the *i*th positive or negative first-order Littrow angle θ_i.

Sources of Uncertainty	Symbol	Type	Uncertainty Value	Probability Distribution	Divisor	Standard Uncertainty u	Sensitivity Coefficient c	$|c|\cdot|u|$
Stability of the angular output	u_{stb}	A	0.06475 arc-second	Gaussian	$\sqrt{200}$	4.579×10^{-3} arc-second	1	4.579×10^{-3} arc-second
Resolution of rotary encoder	u_{res}	B	0.0038 arc-second	Rectangular	$\sqrt{3}$	2.194×10^{-3} arc-second	1	2.194×10^{-3} arc-second
				$u(\theta_i)$				5.077×10^{-3} arc-second

(d) Uncertainty of the light wavelength detected at each Littrow configuration ($u(\lambda i)$)

In the proposed method, the light wavelength λ_{pos_i} or λ_{neg_i} for pitch measurement is evaluated based on Equation (7) as the peak wavelength of the obtained spectrum. Therefore, light intensity fluctuation at each light wavelength and the uncertainty of the reading in the optical spectrum analyzer could contribute to the uncertainty $u(\lambda_i)$ of the light wavelength detected at each Littrow configuration. In this study, a value (7.10×10^{-4} nm) from the previous work by the authors' team [24] was employed as $u(\lambda_i)$.

(e) Uncertainty of the refractive index in air nair ($u(nair)$)

The contribution of n_{air} was then estimated. Since the experiment had been conducted in a laboratory environment, the contributions of the atmospheric pressure, air temperature, and humidity in the room atmospheric range should be considered. It should be noted that a lot of work has been made so far to evaluate the contribution of n_{air}. In this study, taking a value from the literature [25–28], $u(n_{air})$ was estimated to be $\pm 5 \times 10^{-8}$ [26].

Regarding the contribution of each uncertainty source, standard uncertainties u_{Pi} of the pitch P_i obtained by the ith positive or negative first-order Littrow angle can be estimated by the following equations:

$$u_{Pi} = \sqrt{(c(\alpha) \cdot u(\alpha))^2 + (c(\beta) \cdot u(\beta))^2 + (c(\theta_i) \cdot u(\theta_i))^2 + (c(\lambda_i) \cdot u(\lambda_i))^2 + (c(n_{air}) \cdot u(n_{air}))^2} \tag{15}$$

where $c(\alpha)$, $c(\beta)$, $c(\theta_i)$, $c(\lambda_i)$ and $c(n_{air})$ are the sensitivity coefficients that can be described by the following equations:

$$c(\alpha) = \frac{\lambda_i \sin\left(\frac{\alpha-\beta}{2}\right)}{4n_{air}\cos^2\left(\frac{\alpha-\beta}{2}\right)\sin\left(\theta_i + \frac{\beta}{2}\right)} \tag{16}$$

$$c(\beta) = \frac{-\lambda_i\left[\cos\left(\theta_i + \frac{\beta}{2}\right)\cos\left(\frac{\alpha-\beta}{2}\right) + \sin\left(\theta_i + \frac{\beta}{2}\right)\sin\left(\frac{\alpha-\beta}{2}\right)\right]}{4n_{air}\left[\cos\left(\frac{\alpha-\beta}{2}\right)\sin\left(\theta_i + \frac{\beta}{2}\right)\right]^2} \tag{17}$$

$$c(\theta_i) = \frac{-\lambda_i \cos\left(\theta_i + \frac{\beta}{2}\right)}{2n_{air}\cos\left(\frac{\alpha-\beta}{2}\right)\sin^2\left(\theta_i + \frac{\beta}{2}\right)} \tag{18}$$

$$c(\lambda_i) = \frac{1}{2n_{air}\cos\left(\frac{\alpha-\beta}{2}\right)\sin\left(\theta_i + \frac{\beta}{2}\right)} \tag{19}$$

$$c(n_{air}) = \frac{-\lambda_i}{2n_{air}^2\cos\left(\frac{\alpha-\beta}{2}\right)\sin\left(\theta_i + \frac{\beta}{2}\right)} \tag{20}$$

In the above equations, α is the angular misalignment of the measurement laser beam about the Z-axis, β is the angular misalignment value of the measurement laser beam about the X-axis, θ_i is the average Littrow angle, λ_i is the central wavelength, and n_{air} is the refractive index in air. Table 3 summarizes the contribution of each uncertainty source. Based on Equation (15), u_{Pi} was evaluated to be 40.65 pm.

Table 3. Combined uncertainty.

Sources of Uncertainty	Symbol	Standard Uncertainty u	Sensitivity Coefficient c	$|c| \cdot |u|$ pm
α	$u(\alpha)$	3.862×10^{-3} arc-second	9.02×10^{-7} nm/arc-second	3.484×10^{-6}
β	$u(\beta)$	0.187 arc-second	-0.217 nm/arc-second	40.58
θ_i	$u(\theta_i)$	5.077×10^{-3} arc-second	-0.434 nm/arc-second	2.203
λ_i	$u(\lambda_i)$	7.100×10^{-4} nm	1.062	0.754
n_{air}	$u(n_{air})$	5.0×10^{-8}	-833.263 nm	4.166×10^{-2}
Combined uncertainty u_{Pi}				40.65 pm

An expanded uncertainty U of the pitch was thus evaluated to be 81.30 pm ($k = 2$, 95% confidence). From the results shown in Figure 12b, the pitch values were found to distribute in a range from 833.404 ± 0.0372 nm; the experimental results were found to be within the uncertainty value estimated in the above analysis. It should be noted that U estimated in the above analysis is the expanded uncertainty of the pitch by a specific optical mode in the optical frequency comb laser. By using N optical modes ($N = 100$ in Figure 12) observed in experiments, an expanded uncertainty of the pitch measurement by the proposed method becomes $81.30/\sqrt{N}$ pm.

5. Conclusions

For the improvement of the accuracy of pitch measurement of a diffraction grating based on the laser diffraction with an optical frequency comb laser source, major modifications have been made to the optical setup and signal processing. Instead of employing a Fabry-Pérot etalon for the separation of the optical modes in the spectrum of the captured first-order diffracted beams, a weight of mass method has been employed in the signal processing to find out the peak wavelength in the spectrum for the calculation of the grating pitch. In addition, a mathematical model of the pitch measurement in the proposed method with the employment of the optical frequency comb laser has been established while considering misalignments of optical components in the setup. Experiments have been carried out to compensate for the misalignments of the optical components considered in the mathematical model. In addition, the influence of the non-uniform spectrum of the mode-locked femtosecond laser source employed in the developed setup has successfully been reduced through the normalization operations of the light spectra obtained by the detector unit in the setup. Furthermore, by using the developed mathematical model, the theoretical analysis of the uncertainty in the pitch measurement by the proposed method has been carried out based on the guide to the expression of uncertainty in measurement (GUM). Through the theoretical evaluation based on GUM, an expanded uncertainty U of the pitch measurement of a diffraction grating with a nominal pitch density of/1200 mm by using the information from a specific mode has been evaluated to be 81.30 pm ($k = 2.95\%$ confidence). The experimental results were found to be within the uncertainty value estimated in the above analysis. By using N optical modes observed in experiments, an expanded uncertainty of the pitch measurement by the proposed method can be reduced to $81.30/\sqrt{N}$ pm. These results of the theoretical investigation and experiments have demonstrated the feasibility of the proposed pitch measurement with an optical frequency comb laser.

Author Contributions: Conceptualization, W.G. and Y.S.; methodology, D.W.S., L.Q., Y.C., Y.S. and W.G.; software, D.W.S., L.Q. and Y.S.; validation, Y.S., W.G. and H.M.; formal Analysis, D.W.S., L.Q., Y.S. and W.G.; investigation, D.W.S., L.Q., Y.C. and Y.S.; resources, Y.S. and W.G.; data Curation, Y.S. and W.G; Writing—Original Draft Preparation, D.W.S. and Y.S.; Writing—Review and Editing, W.G., Y.S. and E.M.; Visualization, W.G. and Y.S.; Supervision, W.G.; Project Administration, W.G.; Funding Acquisition, W.G., Y.S. and H.M. All authors have read and agreed to the published version of the manuscript.

Funding: This work is supported by the Japan Society for the Promotion of Science (JSPS) 20H00211.

Institutional Review Board Statement: Not applicable.

Informed Consent Statement: Not applicable.

Data Availability Statement: The data presented in this study are available on request from the corresponding author.

Acknowledgments: Dong Wook Shin would like to thank the Graduate Program for Integration of Mechanical Systems (GP-Mech) of Tohoku University, Japan.

Conflicts of Interest: The authors declare no conflict of interest. The funders had no role in the design of the study; in the collection, analyses, or interpretation of data; in the writing of the manuscript or in the decision to publish the results.

References

1. Gao, W.; Kim, S.W.; Bosse, H.; Haitjema, H.; Chen, Y.L.; Lu, X.D.; Knapp, W.; Weckenmann, A.; Estler, W.T.; Kunzmann, H. Measurement technologies for precision positioning. *CIRP Ann.—Manuf. Technol.* **2015**, *64*, 773–796. [CrossRef]
2. Lin, J.; Guan, J.; Wen, F.; Tan, J. High-resolution and wide range displacement measurement based on planar grating. *Opt. Commun.* **2017**, *404*, 132–138. [CrossRef]
3. Ye, W.; Zhang, M.; Zhu, Y.; Wang, L.; Hu, J.; Li, X.; Hu, C. Real-time displacement calculation and offline geometric calibration of the grating interferometer system for ultra-precision wafer stage measurement. *Precis. Eng.* **2019**, *60*, 413–420. [CrossRef]

4. Dai, G.; Koenders, L.; Pohlenz, F.; Dziomba, T.; Danzebrink, H.U. Accurate and traceable calibration of one-dimensional gratings. *Meas. Sci. Technol.* **2005**, *16*, 1241–1249. [CrossRef]

5. Dai, G.; Pohlenz, F.; Dziomba, T.; Xu, M.; Diener, A.; Koenders, L.; Danzebrink, H.U. Accurate and traceable calibration of two-dimensional gratings. *Meas. Sci. Technol.* **2007**, *18*, 415–421. [CrossRef]

6. Misumi, I.; Gonda, S.; Huang, Q.; Keem, T.; Kurosawa, T.; Fujii, A.; Hisata, N.; Yamagishi, T.; Fujimoto, H.; Enjoji, K.; et al. Sub-hundred nanometre pitch measurements using an AFM with differential laser interferometers for designing usable lateral scales. *Meas. Sci. Technol.* **2005**, *16*, 2080–2090. [CrossRef]

7. Chen, J.; Liu, J.; Wang, X.; Zhang, L.; Deng, X.; Cheng, X.; Li, T. Optimization of nano-grating pitch evaluation method based on line edge roughness analysis. *Meas. Sci. Rev.* **2017**, *17*, 264–268. [CrossRef]

8. Brasil, D.A.; Alves, J.A.P.; Pekelsky, J.R. An imaging grating diffractometer for traceable calibration of grating pitch in the range 20 μm to 350 nm. *J. Phys. Conf. Ser.* **2015**, *648*, 012013. [CrossRef]

9. Pekelsky, J.R.; Eves, B.J.; Nistico, P.R.; Decker, J.E. Imaging laser diffractometer for traceable grating pitch calibration. *Meas. Sci. Technol.* **2007**, *18*, 375–383. [CrossRef]

10. Buhr, E.; Michaelis, W.; Diener, A.; Mirandé, W. Multi-wavelength VIS/UV optical diffractometer for high-accuracy calibration of nano-scale pitch standards. *Meas. Sci. Technol.* **2007**, *18*, 667–674. [CrossRef]

11. Sheng, B.; Chen, G.; Huang, Y.; Luo, L. Measurement of grating groove density using multiple diffraction orders and one standard wavelength. *Appl. Opt.* **2018**, *57*, 2514. [CrossRef] [PubMed]

12. Xie, Y.F.; Jia, W.; Zhao, D.; Ye, Z.H.; Sun, P.; Xiang, C.C.; Wang, J.; Zhou, C. Traceable and long-range grating pitch measurement with picometer resolution. *Opt. Commun.* **2020**, *476*, 126316. [CrossRef]

13. Xiong, X.; Matsukuma, H.; Shimizu, Y.; Gao, W. Evaluation of the pitch deviation of a linear scale based on a self-calibration method with a Fizeau interferometer. *Meas. Sci. Technol.* **2020**, *31*, 094002. [CrossRef]

14. Gao, W.; Haitjema, H.; Fang, F.Z.; Leach, R.K.; Cheung, C.F.; Savio, E.; Linares, J.M. On-machine and in-process surface metrology for precision manufacturing. *CIRP Ann.* **2019**, *68*, 843–866. [CrossRef]

15. Cundiff, S.T.; Fortier, T.M.; Ye, J.; Hall, J.L. Carrier-envelope phase stabilization of femtosecond modelocked lasers and direct optical frequency synthesis. In Proceedings of the Conference on Lasers and Electro-Optics, Postconference Technical Digest (IEEE Cat. No.01CH37170), Baltimore, MD, USA, 11 May 2001; p. 130.

16. Leopardi, H.; Davila-Rodriguez, J.; Quinlan, F.; Olson, J.; Sherman, J.A.; Diddams, S.A.; Fortier, T.M. Single-branch Er:fiber frequency comb for precision optical metrology with 10^{-18} fractional instability. *Optica* **2017**, *4*, 879. [CrossRef]

17. Shimizu, Y.; Uehara, K.; Matsukuma, H.; Gao, W. Evaluation of the grating period based on laser diffraction by using a mode-locked femtosecond laser beam. *J. Adv. Mech. Des. Syst. Manuf.* **2018**, *12*, 1–10. [CrossRef]

18. Chen, Y.-L.; Shimizu, Y.; Tamada, J.; Kudo, Y.; Madokoro, S.; Nakamura, K.; Gao, W. Optical frequency domain angle measurement in a femtosecond laser autocollimator. *Opt. Express* **2017**, *25*, 16725. [CrossRef]

19. Korotkov, V.I.; Pulkin, S.A.; Vitushkin, A.L.; Vitushkin, L.F. Laser interferometric diffractometry for measurements of diffraction grating spacing. *Appl. Opt.* **1996**, *35*, 4782. [CrossRef] [PubMed]

20. Decker, J.E.; Eves, B.J.; Pekelsky, J.R.; Douglas, R.J. Evaluation of uncertainty in grating pitch measurement by optical diffraction using Monte Carlo methods. *Meas. Sci. Technol.* **2011**, *22*, 027001. [CrossRef]

21. Shimizu, Y.; Matsukuma, H.; Gao, W. Optical angle sensor technology based on the optical frequency comb laser. *Appl. Sci.* **2020**, *10*, 4047. [CrossRef]

22. Gao, W.; Saito, Y.; Muto, H.; Arai, Y.; Shimizu, Y. A three-axis autocollimator for detection of angular error motions of a precision stage. *CIRP Ann.—Manuf. Technol.* **2011**, *60*, 515–518. [CrossRef]

23. ISO Evaluation of measurement data—Guide to the expression of uncertainty in measurement. *Int. Organ. Stand. Geneva ISBN* **2008**, *50*, 134.

24. Sato, R.; Chen, C.; Matsukuma, H.; Shimizu, Y.; Gao, W. A new signal processing method for a differential chromatic confocal probe with a mode-locked femtosecond laser. *Meas. Sci. Technol.* **2020**, *31*, 094004. [CrossRef]

25. Owens, J.C. Optical Refractive Index of Air: Dependence on Pressure, Temperature and Composition. *Appl. Opt.* **1967**, *6*, 51. [CrossRef] [PubMed]

26. Peck, E.R.; Reeder, K. Dispersion of Air. *J. Opt. Soc. Am.* **1972**, *62*, 958–962. [CrossRef]

27. Ciddor, P.E. Refractive index of air: New equations for the visible and near infrared. *Appl. Opt.* **1996**, *35*, 1566. [CrossRef] [PubMed]

28. Birch, K.P.; Downs, M.J. An updated Edlén equation for the refractive index of air. *Metrologia* **1993**, *30*, 155–162. [CrossRef]

 *applied sciences*

Article

A Real-Time Automated System for Dual-Aperture Common-Path Interferometer Phase-Shifting

Antonio Barcelata-Pinzón [1], Ricardo Iván Álvarez-Tamayo [2,*] and Patricia Prieto-Cortés [3]

[1] Mechatronics Division, Universidad Tecnológica de Puebla, Puebla 72300, Mexico; antonio.barcelata@utpuebla.edu.mx
[2] Faculty of Mechatronics, Electronics, Bionics and Aerospace, Universidad Popular Autónoma del Estado de Puebla, Puebla 72410, Mexico
[3] Faculty of Physics and Mathematics, Universidad Autónoma de Nuevo León, San Nicolás de los Garza 66460, Mexico; patricia.prietocts@uanl.edu.mx
* Correspondence: ricardoivan.alvarez01@upaep.mx

Abstract: We report a novel fully real-time automatized optomechatronic dual-aperture common-path interferometer system for obtaining the phase difference between two interferograms by using the technique of phase-shifting interferometry. A motorized system is used to shift an additional phase transversally to the optical axis by ruling translation. For each high-resolution ruling displacement step of 0.793 μm, an interferogram is recorded by a CCD camera. The phase difference between the two successive recorded interferograms is then automatically calculated by computational self-calibrated algorithms. The proposed device provides more accurate measuring than typically used manual processes. Real-time phase differences are obtained from a robust low-cost optomechatronic system. Analytical calculation of the phase is performed automatically without the requirement of additional or external tools and processes, reducing the significant rework delay. A set of 47 interferograms were captured in real time then recorded and analyzed, obtaining an average phase shifting of 2.483 rad. Analytic explanation and experimental results are presented.

Keywords: generalized phase shifting interferometry; dual-aperture common-path interferometer; real-time optical instrumentation; optomechatronic systems

Citation: Barcelata-Pinzón, A.; Álvarez-Tamayo, R.I.; Prieto-Cortés, P. A Real-Time Automated System for Dual-Aperture Common-Path Interferometer Phase-Shifting. *Appl. Sci.* **2021**, *11*, 7438. https://doi.org/10.3390/app11167438

Academic Editor: Andreas Fischer

Received: 9 July 2021
Accepted: 6 August 2021
Published: 13 August 2021

Publisher's Note: MDPI stays neutral with regard to jurisdictional claims in published maps and institutional affiliations.

Copyright: © 2021 by the authors. Licensee MDPI, Basel, Switzerland. This article is an open access article distributed under the terms and conditions of the Creative Commons Attribution (CC BY) license (https://creativecommons.org/licenses/by/4.0/).

1. Introduction

Mass production through highly automated manufacturing processes has highlighted the need for more efficient precision surface measurement systems and testing processes. In automated manufacturing, the quality of the workpiece surface is highly related to the machining process [1]. Additionally, the manufacturing of technological devices has rapidly migrated to the production of micrometric and nano-scale surfaces [2].

Currently, on-machine in-process measurement provides reliable methodologies to obtain information on machining processes. In-process measurement characteristics require high-speed, high-resolution, real-time measurement, and optical measurement systems provide reliable alternatives to perform measurements and monitoring. Interferometric approaches have been widely used for surface measurement because they can provide noncontact, high-resolution, and high-accuracy measurements [3–5]. Particularly, common-path interferometers such as Fizeau or Twyman–Green have been widely used for ex-situ surface measurement, however, for on-machine measurement it is necessary to reduce the influence of disturbances associated with the process [4,5]. In this sense, it is necessary to develop automated measurement systems, based on optical interferometry with real-time operation, that are immune to disturbances during the measurement of noisy interferograms.

In optical interferometry, phase difference between two interferograms allows physical variable quantities such as displacement [6,7], temperature [8,9], refractive index [9,10], and

---

deformation [11,12], among others, to be obtained. The phase difference can be obtained by applying phase-shifting methods [13–15]. These methods typically consist of introducing an additional phase by shifting either a grating or a mirror depending on the interferometric arrangement configuration [16–18]. The phase-shifting process is typically achieved manually in most cases or by using electro-mechanical actuators [19–21]. In the case of manual shifting, the use of specialized or expensive equipment is not required, however, low accuracy and repeatability of the process is obtained. Alternatively, the use of electromechanical actuators improves the accuracy and linearity of the phase-shifting step spacing; however, the implementation of such systems typically require expensive equipment and specific implementation conditions. In this regard, piezoelectric devices have been widely used to reach millimetric-sized or even nanometric phase-shift linear displacements within interferometric approaches to obtain smaller phase variations [22,23]. Additionally, the interferometric implementation can significantly improve the accuracy when obtaining the phase difference. Particularly, the dual-aperture common-path interferometer (DACPI) is a robust interferometric system that is able to deal with mechanical disturbances better, compared with separated-path interferometers [15,17,24,25]. The DACPI consists of a telecentric, $4f$-Fourier imaging system with two windows in the object plane and a binary ruling as a spatial filter [18].

In DACPI systems, phase stepping is usually obtained by transverse translation of a Ronchi ruling placed at the Fourier plane, where the translation is typically performed manually [17]. Moreover, the phase difference obtained by a DACPI requires external digital processing, which includes the use of algorithms, usually performed separately from the capture stage. In this regard, the main disadvantages in systems such as DACPI is the significant time increase in phase calculating and related miscalibration issues.

In this paper, we present an automated DACPI system for obtaining the phase difference between the probe and the reference window in real time. Furthermore, the obtaining of the phase extraction is carried out by means of a self-calibrated algorithm [26,27], without the requirement of equal spaced consecutively captured interference patterns [13,14,28]. The computational process for the phase-difference calculation is performed in real time at the image plane. The proposed system is a reliable alternative to improve the accuracy of the phase difference calculation, incorporate low-cost elements, and use a homemade ruling translation system.

2. Operation Principle of a DACPI

A DACPI is a telecentric $4f$-imaging system formed by a couple of windows in the object plane and a Ronchi ruling used as spatial filter in the Fourier plane, as it is shown in the schematic of Figure 1. One window acts as a reference arm where a phase object is placed, whereas the other one is the probe arm. In the first section (window stage), a collimated laser beam with wavelength λ propagates through the windows A and B with dimensions determined by a_w and b_w placed in the object plane. In order to match the interference condition, the distance between window centers is $x_0 \leq \lambda f / u_p$, where u_p is the ruling period and f is the focal length, determined by the lenses L_1 and L_2.

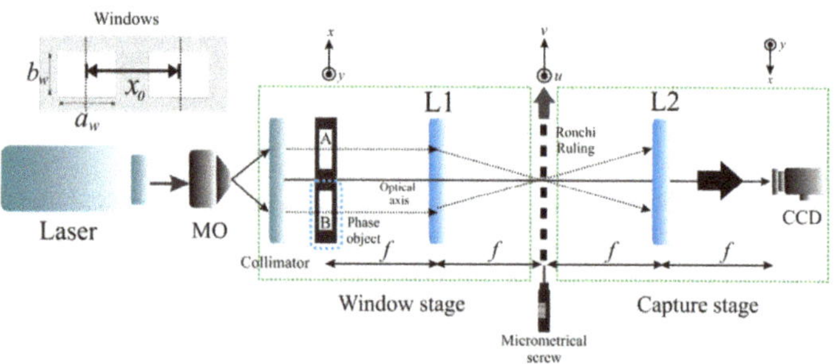

Figure 1. Schematic of a DACPI; MO: Microscope Objective, CCD: Charge-Coupled Device camera.

The field from the object plane is propagated through L_1, then, in the ruling plane, the field is formed by the interfered diffraction orders from both window fields. Considering the impulse response of the spatial filter, the rectangular function for the windows at the image plane (capture stage) is expressed by $w(x, y) = \text{rect}(x/a_w)\text{rect}(y/b_w)$, and the phase difference between the probe and the reference arm is $\phi(x, y) = \phi(x, y)_R - \phi(x, y)_P$. The interference pattern due to the superposition of fields from the 0th order of the probe arm and the 1st order of reference arm can be described as in [18,29]:

$$I_{0j}(x, y) = \frac{1}{\pi^2} A_R{}^2 + \frac{1}{4} A_P{}^2 + \frac{1}{\pi} A_R A_P \cos(\phi(x, y) - \alpha_j), \tag{1}$$

where A_R and A_P are the output amplitudes of the beams from the windows and α_j is the phase added to generate j known phase steps. Therefore, the phase-shifting interferometry (PSI) is based on the generation of additional phases α_j and the capture of intensities for each corresponding phase. In a DACPI, this additional phase can be achieved by transversal ruling translation on the Fourier plane, where the amount of phase added is given by [16,17]:

$$\alpha_j = 2\pi \frac{u_d}{u_p}, \tag{2}$$

where u_p is the ruling period and u_d is the linear displacement of the ruling. As the values of u_p and u_d are known, it is possible to generate N number of steps to apply a self-calibrated algorithm in order to recover the phase $\phi(x, y)$.

3. Real-Time Automated Phase-Difference Retrieving from a DACPI

Typically, in sensing DACPI applications, the determination of $\phi(x, y)$ allows the estimation of a physical, chemical or biological quantity. Here, a measurable quantity from a transparent object placed at the probe arm can be related to the phase difference between the light beams from both arms [15,29–31]. For practical applications, it is convenient to obtain an immediate value of $\phi(x, y)$ to achieve real-time monitoring of the measured quantity, which also allows fast and precise calibration of the system. From Equation (2) it can be clearly noticed that the precision of the phase amount directly depends on the control of the transverse translation of the ruling, since u_p is a fixed parameter that depends on the ruling fabrication characteristics. As a consequence, u_d is the only parameter that can be tailored in order to obtain phase steps for specific applications. In most of the cases, the linear displacement of the ruling is manually performed, which increases the measurement inconsistencies due to human error. The automation of the u_d parameter by using an electromechanical system then improves the reliability and repeatability of the system.

Figure 2 shows a schematic diagram of the automation system. In order to generate interference fringes, a transparent object is placed in the probe window B. The interfering fields, denoted in Equation (1), generate interferograms described by fringe patterns $I_j(x, y)$ that are produced by the comparison between the reference and the probe wavefronts. The intensity of the fringe pattern is captured by the CCD camera and digitally recorded as 8-bit grayscale images. Then, by using controlled electromechanical rotation on the micrometrical screw, the ruling is linearly displaced (in the direction of the x-axis in Figure 1) generating a phase change α_{j+1} in the interference field. Therefore, a new and consecutive grayscale pattern $I_{j+1}(x, y)$ is obtained and then recorded. Subsequently, it can obtain as many patterns $I_j(x, y)$ as there are induced phase changes α_j.

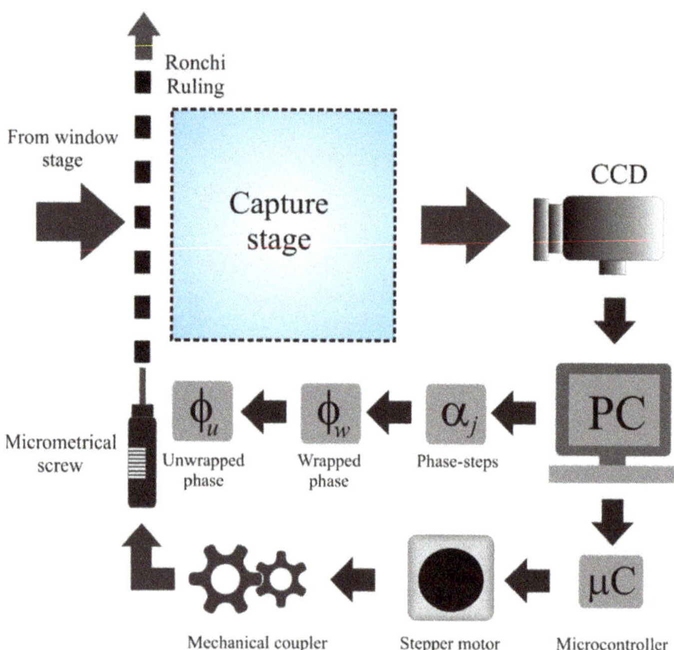

Figure 2. Schematic of the automation system arrangement.

A microcontroller (PIC16F877A) receives codified instructions of the image capture and processing of the jth interferogram from the LabVIEW platform. Then, the microcontroller, aided by a power stage, delivers the 4-bit impulse sequence required for a unipolar stepper motor (Nema17) operation (see Figure 3a). The stepper motor is coupled to a PLA 3D printed spur gear system to rotate a micrometric screw (Mitutoyo Head Series 149) of an optical translation stage, as it is shown in Figure 3b. The gear transmission produces a linear displacement of a Ronchi ruling with a period of 500 lines/mm $\left(u_p = 2\ \mu\text{m}\right)$ mounted on the translation stage. In order to meet the interference condition in the object plane, the separation between the windows was set to $x_0 = 30$ mm, considering a He–Ne laser with $\lambda = 632.8$ nm and focal length $f = 100$ mm. The micrometer screw had a nominal resolution of 0.001 inches per division (25.4 μm), then, without a gear transmission system, a full turn is completed with a linear displacement of 0.025 inches (635 μm). With a stepper motor resolution of 1.8°, the ruling is transversally displaced 3.175 μm per motor step. According to Equation (2), the ruling displacement generates an aggregate phase of $\alpha = 2\pi(3.175\ \mu\text{m})/2\ \mu\text{m} \approx 10$ rad to the interference pattern, which exceeds the range of $(0, 2\pi)$ in which the algorithms can calculate the phase. Therefore, a gear transmission system coupled to the micrometric screw allows the decreasing of the ruling linear displacement to meet the range from 0 to 2π rad. Figure 3c shows the designed 1:4 gear transmission system, which decreases the linear displacement of the ruling to $u_d = 0.793$ μm, allowing theoretical phase shifts of $\alpha \approx 2.493$ rad.

It is worth mentioning that a commercially available piezoelectric optical translation system, which includes a driver and a coupling device, is approximately 1200% more expensive compared with our system based on a stepper motor system. Although a piezoelectric device can reach higher angular displacement resolution, our setup represents a reliable low-cost approach for real-time automatic phase recovery by phase shifting from a dual-aperture common-path interferometer.

Furthermore, the integration through a widely distributed platform, such as LabView, allows standard low-cost access to users interested in the proposed system.

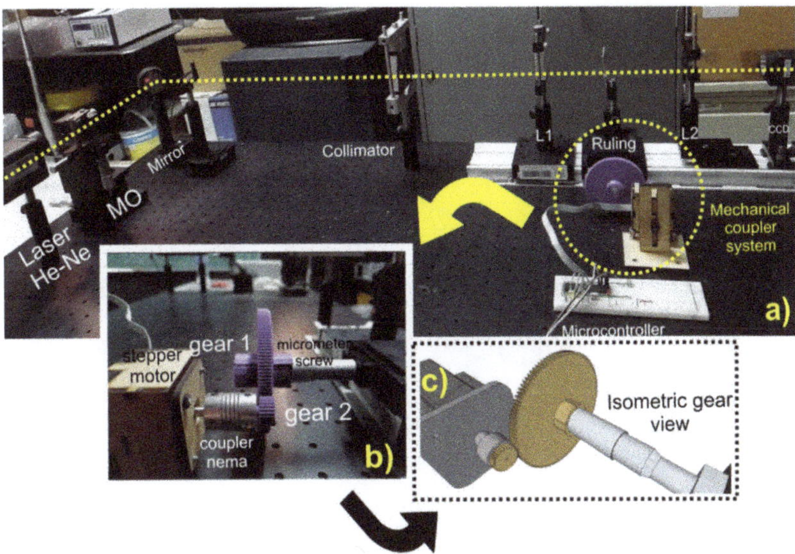

Figure 3. An automated DACPI system: (**a**) implementation of the system, (**b**) detail of the gear transmission system, and (**c**) the designed spur gear transmission.

Generally, phase-shifting techniques retrieve the object phase $\phi(x,y)$ by solving an $N \times 3$ system of equations from equal phase steps and specific phase-shifter calibration [32]. Particularly, generalized phase-shifting interferometry (GPSI) does not require known and equal phase steps, however a specific phase-shifter calibration is required. As recently reported, self-calibration generalized phase-shifting interferometry (SGPSI) provides the possibility of unequal and unknown phase steps without the requirement of an exhaustive calibrating process to retrieve the object phase [26,27]. In this case, SGPSI advantages allow a straight forward calculation of the phase-step α introduced by linear displacement of the ruling. Digital grayscale pattern images are converted into an $m \times n$ matrix in order to apply the SGPSI technique and obtain the phase differences α between recorded interferograms using [27]:

$$\cos \tfrac{1}{2}\alpha_1 = \frac{\mathrm{tr}(q^{\mathrm{T}}r)}{\sqrt{\mathrm{tr}(q^{\mathrm{T}}q)\mathrm{tr}(r^{\mathrm{T}}r)}}$$

$$\cos \tfrac{1}{2}\alpha_2 = \frac{\mathrm{tr}(p^{\mathrm{T}}q)}{\sqrt{\mathrm{tr}(p^{\mathrm{T}}p)\mathrm{tr}(q^{\mathrm{T}}q)}} \tag{3}$$

$\mathrm{tr}(\cdot)$ and $(\cdot)^{\mathrm{T}}$ are the trace and transpose operators respectively; the restrictions are $\alpha_0 = 0, \alpha_1 \in (0, \alpha_2)$ and $\alpha_2 \in (0, 2\pi)$, $p = I_0 - I_1$, $q = I_1 - I_2$, and $r = I_0 - I_2$. Then, the phase of the probe arm $\phi(x,y)$ is obtained by:

$$\tan \phi = \frac{I_2 - I_1 + (I_0 - I_2)\cos\alpha_1 - (I_0 - I_1)\cos\alpha_2}{(I_0 - I_2)\sin\alpha_1 - (I_0 - I_1)\sin\alpha_2} \tag{4}$$

The block diagram of the SGPSI algorithm used to obtain $\phi(x,y)$, including the automation, is shown in Figure 4. The automation of the phase retrieving by the proposed DACPI system was incorporated into the implementation in LabView, used for the capture of the interferograms. Equation (4) retrieves the original phase function, for this purpose the SGPSI requires a minimum of 3 phase-shifted interference patterns in order to achieve the values of α_j. Each pattern is captured by a CCD camera from sequential motor steps, and then, converted into grayscale 2D $m \times n$ arrays: I_0, I_1, and I_2. The SGPSI algorithm is performed using the values of p, q, and r, obtained from the I_0, I_1, and I_2 interactions. Then,

α_1 and α_2 are calculated ($\alpha_0 = 0$). Next, the phase ϕ_w is calculated, wrapped for a period 2π or integer multiples of 2π, which leads to a discontinuous function in the solution. In this regard, several techniques known as unwrapping phase methods are used to eliminate such discontinuities to obtain a smoothed function of the phase by adding 2π or multiples of 2π. LabView libraries includes the virtual instrument (VI) Unwrap Phase VI, which allows 1D phase unwrapping. In order to simplify the design of the algorithm, the VI was used and adapted for the 2D unwrapping of ϕ_w. The unwrapped phase ϕ is then obtained.

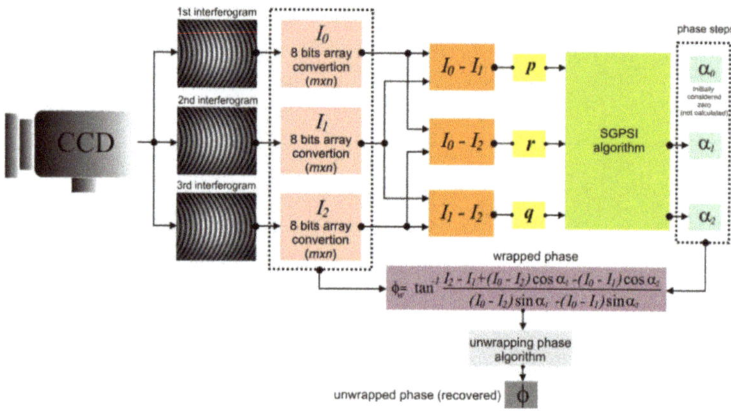

Figure 4. Block diagram of the implemented SGPSI algorithm for the phase obtaining.

4. Results and Discussion

Once the mechatronic system was implemented, computational tests were performed. The algorithm was proved using three simulated phase-shifted interference patterns labeled as (a), (b), and (c) in the LabView front panel, shown in Figure 5. The original function from the simulated phase steps of the interferograms was shown as reference in the plotting labeled as (e). Then, the proposed retrieving phase algorithm calculated the values of p, q, r from the simulated interferograms to obtain the phase step values of α. Table 1 shows the comparison between the proposed and calculated phase step values α_1 and α_2 and the absolute value of the error. Next, the algorithm calculated the wrapped phase function ϕ_w, observed in the plot (d). Finally, the unwrapped phase ϕ_u is obtained, as it is shown in the plot labeled as (f) in the front panel image, as it is displayed to the user. The plot (f) corresponds to the retrieved 2D unwrapped phase function obtained from the algorithm, which can be compared with the 2D original function of the plot (e). This comparison is shown in plot (g) as the subtraction (error) between them (red curve). Additionally, the plot labeled as (h) shows the calculated 3D phase function. As can be observed in Table 1, the non-zero error is attributed to the limitation of the used SPGSI algorithm, which required a high number of interference fringes to reduce the error [26]. In the proposed system, the error is considered acceptable due to its advantages compared with the reported systems. In our approach, the proposed SPGSI algorithm exhibited noise immunity which allowed the calculation of the phase from the experimental interferograms, avoiding undergoing filtering processes [33]. In addition, equal spacing displacement is not required from the obtained interferograms and the needless special calibration procedures.

Afterwards, experimental results were obtained. In order to generate the interference fringes, transparent acetate was used as the phase object placed in the probe window B of the object plane. It is worth noting that for experimental results, the angular displacements generated in each motor step are not equal. However, for the implemented algorithm, the equal steps condition is not mandatory since the SGPSI is capable of achieving phase shifts, even if these steps are unequal. Figure 6 shows a set of four CCD consecutive interference pattern captures. The images were arranged in two groups of three patterns. The first group, corresponding to the initial phase steps, includes the first three images whose intensities are I_0, I_1, and I_2. The second group stands for the new phase for each step calculation and includes the second, third and, fourth images with intensities I_1, I_2, and I_3. Then, the phase function ϕ_w is obtained by calculating the values of α_1 and α_2 with the corresponding phase difference between each pair of interferograms.

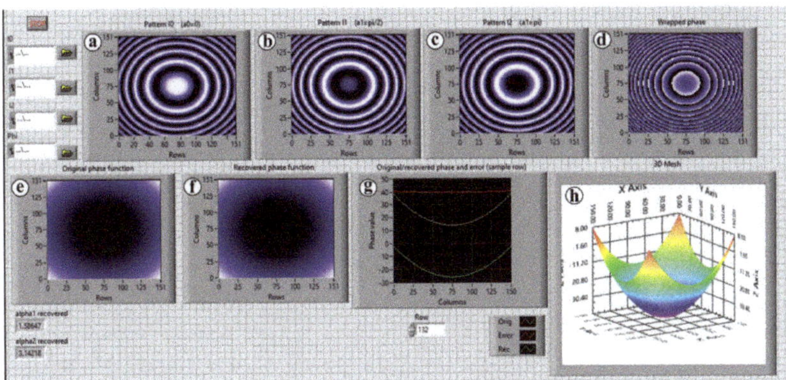

Figure 5. Front panel of LabView for the implemented algorithm with simulated interferograms: (**a**) simulated pattern without phase shift, (**b**) simulated pattern with phase shift of $\pi/2$ rad, (**c**) simulated pattern with phase shift of π rad, (**d**) calculated wrapped phase function, (**e**) original 2D phase function, (**f**) recovered (unwrapped) 2D phase function, (**g**) difference between the original and the recovered phase, and (**h**) recovered 3D phase plot.

Table 1. Comparison between the simulated and the calculated values for the phase shifts.

Simulated Value	Calculated Value	Error (Absolute Value)
$\alpha_0 = 0$	$\alpha_0 = 0$	-
$\alpha_1 = 1.57079$	$\alpha_1 = 1.586469$	0.015
$\alpha_2 = 3.14159$	$\alpha_2 = 3.142185$	0.0005

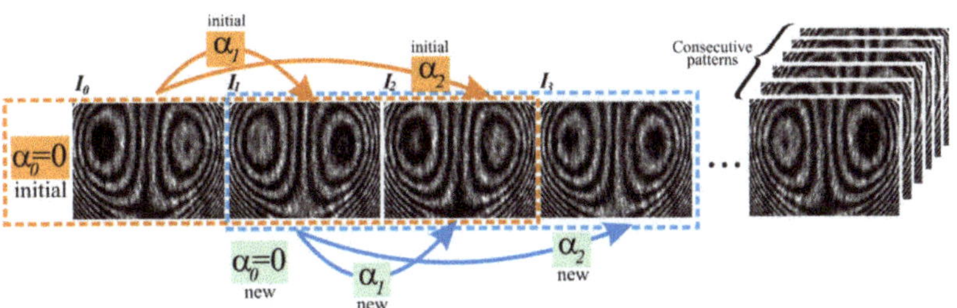

Figure 6. Experimental pattern captures.

As a result of evaluating the set of experimental patterns, the wrapped and unwrapped phase is calculated from the obtained phase steps for each motor step. Figure 7 shows the front panel of LabView for the implemented algorithm. The plots labeled as (a), (b), and (c) correspond to the calculated phase shift for each step. The plot (d) corresponds to the calculated wrapped phase. The 2D and 3D plots for the retrieved phase are labeled as (e) and (f), respectively. As it can be observed, by using three consecutive experimental patterns it is possible to automatically obtain the phase function, without reprocessing the data.

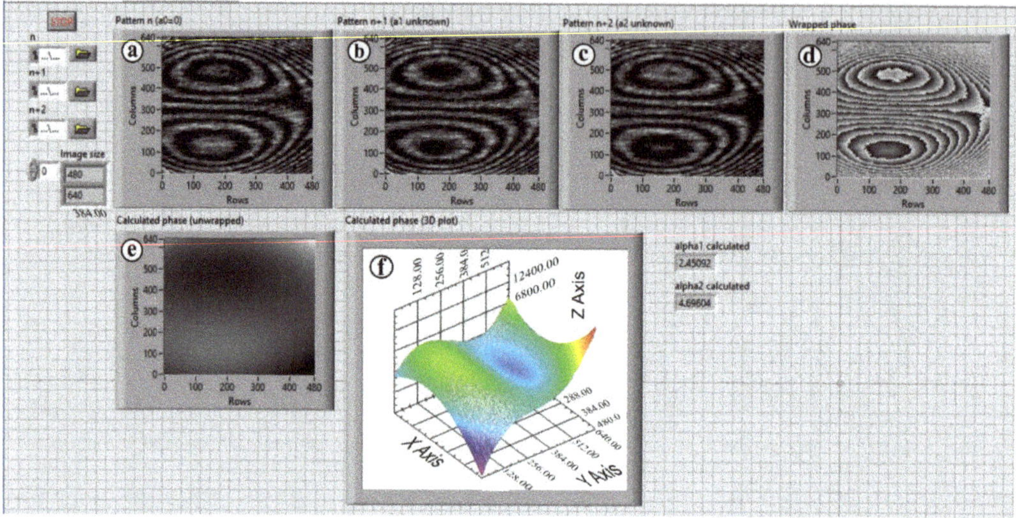

Figure 7. Experimental results: (**a–c**) phase-shifted consecutive experimental interferograms, (**d**) wrapped phase, (**e**) 2D plot unwrapped phase, and (**f**) 3D plot unwrapped phase.

The results of the characterization of the phase shift introduced by each stepper motor rotation step are shown in Figure 8. The measurement was obtained from 47 consecutive experimental pattern captures. The calculated values of α from the consecutive images recorded were obtained by using the aforementioned algorithm. The first set of results was computed with the obtained interferograms from 1 to 3, the second set of results was computed with the obtained interferograms from 2 to 4, and so on to complete all 46 calculations of the phase step. The theoretical phase step value calculated with Equation (2) is $\alpha_T = 2.493$ rad (blue line), whereas the average value of the experimental phase step of $\bar{\alpha} = 2.483$ rad (green line) was obtained from the calculated values of α from the consecutive recorded interferograms. The bias between the theoretical and experimental values is $|\alpha_T - \alpha| = 0.01$ rad. The average standard deviation of the total result is of $\sigma = 0.197$. Finally, it is possible to calculate the average value with the uncertainty percentage as $\bar{\alpha} = 2.483$ rad $\pm 8\%$, which is comparable to the result obtained by manual transversal displacement of the ruling in a DACPI system but with measurements obtained over half a step of the reported values [29]. It is important to mention that because the algorithm is capable of evaluating phase changes with non-uniform shifts, the system was able to obtain the phase function with complex implementation, high-precision devices, and special calibration.

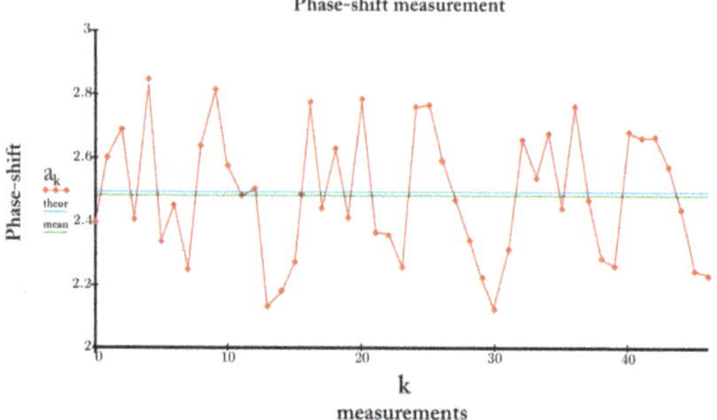

Figure 8. Experimental estimation of error in the phase step.

5. Conclusions

In this paper, we demonstrated the reliable operation of an automated DACPI system with real-time obtaining of the phase difference between the probe and reference windows. The phase extraction was achieved by a self-calibrated algorithm without the requirement of equally spaced displacements.

As a result of the experimental evaluation, it is demonstrated that the implementation of a real-time automatic system capable of transversely shifting a ruling to the optical axis in a dual-aperture common path interferometer is possible. The transverse displacement of the ruling generates a phase change between one interferogram and another, which the system is able to calculate using elements without requirement of extensive calibration. Additionally, the proposed system calculates the unwrapped phase and immediately displays it on a user-friendly platform, avoiding rework and increasing the precision compared to manually operated systems. From a set of 47 interferograms an average motor step phase difference of 2.483 rad was obtained, with a difference of 0.01 rad with respect to the theoretical value.

The proposed automated system is a reliable option to improve the accuracy when calculating the phase difference by using low–cost elements and a homemade ruling translation.

Author Contributions: A.B.-P. and R.I.Á.-T. conceived and designed the experiments; P.P.-C. contributed with data analysis; A.B.-P. designed the computational system to carry out the experiment; R.I.Á.-T. and P.P.-C. performed the simulation; A.B.-P. wrote the paper; R.I.Á.-T. reviewed the paper and P.P.-C. edited the paper. All authors have read and agreed to the published version of the manuscript.

Funding: This research was funded by Universidad Popular Autónoma del Estado de Puebla and Universidad Tecnológica de Puebla.

Institutional Review Board Statement: Not applicable.

Informed Consent Statement: Not applicable.

Acknowledgments: We thank the Optomechatronics group UTPUE-CA-13 and Mechatronics Division of Universidad Tecnológica de Puebla. R.I.Á.-T. wants to thank to Fondo de Investigación UPAEP 2020. P.P.-C. was supported by CONACyT postgraduate scholarship grant 160540. LabView software was used under license granted for Universidad Tecnológica de Puebla.

Conflicts of Interest: The authors declare no conflict of interest.

References

1. Dawei, T.U. In-process sensor for surface profile measurement applying a common-mode rejection technique. *Opt. Laser Technol.* **1995**, *27*, 351–353. [CrossRef]
2. Gao, F. *Optical Interferometry*; Banishev, A.A., Bhowmick, M., Wang, J., Eds.; IntechOpen: London, UK, 2017; Volume 1, p. 42.
3. Meeß, R.; Dontsov, D.; Langlotz, E. Interferometric device for the in-process measurement of diameter variation in the manufacture of ultraprecise spheres. *Meas. Sci. Technol.* **2021**, *32*, 074004. [CrossRef]
4. Kohno, T.; Matsumoto, D.; Yazawa, T.; Uda, Y. Radial shearing interferometer for in-process measurement of diamond turning. *Opt. Eng.* **2000**, *39*, 2696–2699.
5. Tomlinson, R.; Coupland, J.M.; Petzing, J. Synthetic aperture interferometry: In-process measurement of aspheric optics. *Appl. Opt.* **2003**, *42*, 701–707. [CrossRef]
6. Norgia, M.; Donati, S.A. Displacement-Measuring Instrument Utilizing Self-Mixing Interferometry. *IEEE Trans. Instrum. Meas.* **2003**, *52*, 1765–1770. [CrossRef]
7. Kholkin, A.L.; Wütchrich, C.; Taylor, D.V. Setter N. Interferometric measurements of electric field-induced displacements in piezoelectric thin films. *Rev. Sci. Instrum.* **1996**, *67*, 1935–1941. [CrossRef]
8. Shakher, C.; Nirala, A.K. Measurement of temperature using speckle shearing interferometry. *Appl. Opt.* **1994**, *33*, 2125–2127. [CrossRef]
9. Sharma, S.; Sheoran, G.; Shakher, C. Digital holographic interferometry for measurement of temperature in axisymmetric flames. *Appl. Opt.* **2012**, *51*, 3228–3235. [CrossRef]
10. Chiu, M.-H.; Lee, J.-Y.; Su, D.-C. Complex refractive-index measurement based on Fresnel's equations and the uses of heterodyne interferometry. *Appl. Opt.* **1999**, *38*, 4047–4052. [CrossRef] [PubMed]
11. Moore, A.J.; Hand, D.P.; Barton, J.S.; Jones, J.D.C. Transient deformation measurement with electronic speckle pattern interferometry and a high-speed camera. *Appl. Opt.* **1999**, *38*, 1159–1162. [CrossRef]

12. Jacquot, P. Speckle interferometry: A review of the principal methods in use for experimental mechanics applications. *Strain* **2008**, *44*, 57–69. [CrossRef]
13. Kinnstaetter, K.; Lohmann, A.W.; Schwider, J.; Streibl, N. Accuracy of phase shifting interferometry. *Appl. Opt.* **1988**, *27*, 5082–5089. [CrossRef] [PubMed]
14. Joenathan, C. Phase-measuring interferometry: New methods and error analysis. *Appl. Opt.* **1994**, *33*, 4147–4155. [CrossRef]
15. Meneses Fabian, C.; Rodriguez Zurita, G.; Arrizón, V. Optical tomography of transparent objects with phase-shifting interferometry and stepwise-shifted Ronchi ruling. *J. Opt. Soc. Am. A* **2006**, *23*, 298–305. [CrossRef]
16. Meneses Fabian, C.; Rodriguez Zurita, G. Carrier fringes in the two-aperture common-path interferometer. *Opt. Lett.* **2011**, *36*, 642–644. [CrossRef] [PubMed]
17. Meneses Fabian, C.; Rodriguez Zurita, G.; Vazquez Castillo, J.F.; Robledo Sanchez, C.; Arrizón, V. Common-path phase-shifting interferometer with binary grating. *Opt. Commun.* **2006**, *264*, 13–17. [CrossRef]
18. Arrizón, V.; Sánchez-de-la-Llave, D. Common-path interferometry with one-dimensional periodic filters. *Opt. Lett.* **2004**, *29*, 141–143. [CrossRef]
19. Bruno, L.; Poggialini, A.; Felice, G. Design and calibration of a piezoelectric actuator for interferometric applications. *Opt. Lasers Eng.* **2007**, *45*, 1148–1156. [CrossRef]
20. Nguyen, T.D.; Duong, Q.A.; Higuchi, M.; Vu, T.T.; Wei, D.; Aketagawa, M. 19-Picometer Mechanical Step Displacement Measurement Using Heterodyne Interferometer With Phase-Locked Loop and Piezoelectric Driving Flexure-Stage. *Sens. Actuators A Phys.* **2020**, *304*, 1–11. [CrossRef]
21. Sokkar, T.Z.N.; El-Farahaty, K.A.; El-Bakary, M.A.; Omar, E.Z.; Hamza, A.A. Optical birefringence and molecular orientation of crazed fibres utilizing the phase shifting interferometric technique. *Opt. Laser Technol.* **2017**, *94*, 208–216. [CrossRef]
22. Chassagne, L.; Topcu, S.; Alayli, Y.; Juncar, P. Highly accurate positioning control method for piezoelectric actuators based on phase-shifting optoelectronics. *Meas. Sci. Technol.* **2005**, *16*, 1771–1777. [CrossRef]
23. Connelly, M.J.; Galeti, J.H.; Kitano, C. Piezoelectric mirror shifter transfer function measurement, modelling, and analysis using feedback based synthetic-heterodyne Michelson interferometry. *OSA Contin.* **2020**, *3*, 3424–3432. [CrossRef]
24. Shan, M.; Hao, B.; Zhong, Z.; Diao, M.; Zhang, Y. Parallel two-step spatial carrier phase-shifting common-path interferometer with a Ronchi grating outside the Fourier plane. *Opt. Express* **2013**, *21*, 2126–2132. [CrossRef] [PubMed]
25. Mico, V.; Zalevsky, Z.; García, J. Superresolution optical system by common-path interferometry. *Opt. Express* **2006**, *14*, 5168–5177. [CrossRef]
26. Gomez Conde, J.C.; Meneses Fabian, C. Real-time measurements of phase steps out-of-range ($0,2\pi$) by a dynamic self-calibrating generalized phase-shifting algorithm. *Opt. Lasers Eng.* **2021**, *140*, 1–8. [CrossRef]
27. Gomez Conde, J.C.; Meneses Fabian, C. Real-time phase step measurement using the volume enclosed by a surface algorithm in self-calibrating phase-shifting interferometry. *Meas. J. Int. Meas. Confed.* **2020**, *153*, 1–9. [CrossRef]
28. Creath, K. Phase-shifting speckle interferometry. *Appl. Opt.* **1985**, *24*, 3053–3058. [CrossRef] [PubMed]
29. Barcelata Pinzon, A.; Meneses Fabian, C.; Moreno Alvarez, L.; Pastrana Sanchez, R. Common-path speckle interferometer for phase objects studies. *Opt. Commun.* **2013**, *304*, 153–157. [CrossRef]
30. Lopez Ortiz, B.; Toto Arellano, N.I.; Flores Muñoz, V.H.; Martínez García, A.; García Lechuga, L.; Martínez Domínguez, J.A. Phase profile analysis of transparent objects through the use of a two windows interferometer based on a one beam splitter configuration. *Optik* **2014**, *125*, 7227–7230. [CrossRef]
31. Preda, F.; Perri, A.; Réhault, J.; Dutta, B.; Helbing, J.; Cerullo, G.; Polli, D. Time-domain measurement of optical activity by an ultrastable common-path interferometer. *Opt. Lett.* **2018**, *43*, 1882–1885. [CrossRef] [PubMed]
32. Hipp, M.; Woisetschläger, J.; Reiterer, P.; Neger, T. Digital evaluation of interferograms. *Meas. J. Int. Meas. Confed.* **2004**, *36*, 53–66. [CrossRef]
33. Meneses-Fabian, C. Self-calibrating generalized phase-shifting interferometry of three phase-steps based on geometric concept of volume enclosed by a surface. *J. Opt.* **2016**, *18*, 1–12. [CrossRef]

Article

Mirau-Based CSI with Oscillating Reference Mirror for Vibration Compensation in In-Process Applications

Hüseyin Serbes *, Pascal Gollor, Sebastian Hagemeier and Peter Lehmann

Measurement Technology, Department of Electrical Enginering and Computer Science, University of Kassel, Wilhelmshoeher Allee 71, 34121 Kassel, Germany; pascal.gollor@uni-kassel.de (P.G.); sebastian.hagemeier@uni-kassel.de (S.H.); p.lehmann@uni-kassel.de (P.L.)
* Correspondence: h.serbes@uni-kassel.de

Abstract: We present a Mirau-type coherence scanning interferometer (CSI) with an oscillating reference mirror and an integrated interferometric distance sensor (IDS) sharing the optical path with the CSI. The IDS works simultaneously with the CSI and measures the distance changes during the depth scanning process with high temporal resolution. The additional information acquired by the IDS is used to correct the CSI data disturbed by unwanted distance changes due to environmental vibrations subsequent to the measurement. Due to the fixed reference mirror in commercial Mirau objectives, a Mirau attachment (MA) comprising an oscillating reference mirror is designed and built. Compared to our previous systems based on the Michelson and the Linnik interferometer, the MA represents a novel solution that completes the range of possible applications. Due to its advantages, the Mirau setup is the preferred and most frequently used interferometer type in industry. Therefore, the industrial use is ensured by this development. We investigate the functioning of the system and the capability of the vibration compensation by several measurements on various surface topographies.

Keywords: coherence scanning interferometry; in-process application; Mirau interferometer; vibration compensation; interferometric distance sensor; optical path length modulation; oscillating reference mirror

Citation: Serbes, H.; Gollor, P.; Hagemeier, S.; Lehmann, P. Mirau-Based CSI with Oscillating Reference Mirror for Vibration Compensation in In-Process Applications. *Appl. Sci.* **2021**, *11*, 9642. https://doi.org/10.3390/app11209642

Academic Editor: Andreas Fischer

Received: 4 September 2021
Accepted: 12 October 2021
Published: 15 October 2021

Publisher's Note: MDPI stays neutral with regard to jurisdictional claims in published maps and institutional affiliations.

Copyright: © 2021 by the authors. Licensee MDPI, Basel, Switzerland. This article is an open access article distributed under the terms and conditions of the Creative Commons Attribution (CC BY) license (https://creativecommons.org/licenses/by/4.0/).

1. Introduction

Optical topography sensors such as scanning confocal microscopes (SCM) [1] and coherence scanning interferometers (CSI) [2,3] are well-established measuring instruments in industry and research for the contactless inspection of surface topographies with height structures in the micro- and nanometer range. However, these topography sensors are sensitive to external disturbances, which distort the corresponding controlled depth scanning procedure. Therefore, a close-to-machine employment is not possible with conventional SCM and CSI instruments. As a consequence, in industrial applications, samples are investigated by these optical sensors under highly controlled conditions on an anti-vibration table.

One solution for full-field in situ measurements provides fast measuring methods such as single or dual shot approaches [4–6]. However, as the depth of field is required to cover the whole height range of the specimen, the lateral resolution and axial accuracy are usually worse compared to CSI, e.g., in case of the approach presented in [5] the lateral resolution is specified by 30 μm for an NA of 0.28 and thus exceeds the Rayleigh resolution of the system. Another approach is given by the usage of an additional laser interferometer besides the main measuring system, the output of which is used to actively compensate for deflections caused by environmental vibrations [7–10]. Schäfer et al. [9] report a setup where vibrations are measured with respect to the measuring table of the specimen using a quadrature interferometer. The quadrature signal is used to move the entire CSI contrarily to the deflections caused by the vibration using a piezo driven stage. In contrast, Teale et al. [10] present a CSI with a vibration compensation for in situ depth measurements in a deep reactive ion etcher. Besides the light of low temporal coherence used for CSI, a laser beam propagates through the Michelson interferometer in order to detect vibrations. The output of the laser interferometer is used in a closed-loop with a piezo element, which actively compensates for vibrations by moving the reference mirror. Besides these active vibration compensation techniques, passive approaches have also been demonstrated to be suitable for vibration compensation. The recorded interference signals may be postprocessed, as shown by Deck [11] for a phase shifting interferometer, where a so-called phase-error pattern is used to remove deviations caused by vibrations. In order to reduce the effect of noise, Kiselev et al. [12] correlate the measured interferogram with a nominal one.

Another promising approach for passive vibration compensation in CSI is the use of an interferometric distance sensor (IDS) integrated in the optical path of the CSI setup. The changed distance between CSI and surface under investigation caused by a depth scan is measured by the IDS simultaneously to the CSI measurement. This also enables the detection of vibrations, which are represented by an additional change in path length between CSI and specimen. The interference signal obtained by the IDS provides distance measuring values at a high frequency. These are used to correct the CSI datasets subsequent to the measurement. This method is introduced by Tereschenko et al. and is utilized in a Michelson [13,14] as well as in a Linnik [15] interferometer setup. The working principle of the IDS is based on an optical path length modulation (OPLM) due to a sinusoidal oscillation of the used reference mirror. If this OPLM is fast enough, distance changes can be monitored with high temporal resolution. Depending on the oscillation frequency of the reference mirror, high measurement speeds of up to 116 kHz can be achieved [16]. This enables the detection of vibrations comprising ultrasonic frequencies. An early report of this approach is given by Sasaki and Okazaki [17]. The method is also used in several sensor constellations, which are adapt to other areas of application as reported in [16,18–22].

In contrast to the two interferometer arrangements described by Tereschenko et al. [14,15] Mirau-type CSI instruments are usually preferred in industrial and other practical applications due to their advantages, namely, a sufficient numerical aperture (NA), a compact design, as well as low dispersion distortions. Furthermore, Mirau systems encompass a simple handling due to less demanding adjustment compared to, e.g., a Linnik setup. Moreover, significantly lower system costs result from the configuration with only one microscope objective lens [23]. Therefore, extending the range of application and especially making the method accessible for industrial use requires the transfer of the previously described vibration compensation technique to a Mirau CSI. However, due to the need of a sinusoidally oscillating reference mirror, this method is not directly applicable to conventional Mirau-type CSI system. To overcome this, and to make the method described above suitable for use with a Mirau interferometer, a self-designed and build Mirau setup with oscillating reference mirror is presented in this article. Therefore, a microscope objective lens providing a long working distance and a glass thickness compensation is used and extended to a Mirau interferometer by applying a Mirau attachment (MA). The MA consists inter alia of the oscillating reference mirror that enables the IDS to share the optical path with the CSI.

The sensor setup is introduced in Section 2 with focus on the Mirau extension. Afterwards, the IDS signal processing and correction of the CSI signals is described in Section 3. In Section 4, we present measurement results to validate the transfer characteristics and vibration compensation capabilities of our Mirau CSI on several specimens. A conclusion is drawn in Section 5 and an outlook to further investigations will be given.

2. Sensor Configuration

The proposed sensor setup is based on a Mirau CSI. The temporally low coherent light emitted by an LED and the coherent infrared laser beam with a wavelength of 850 nm emitted by a diode laser are combined by a cold light reflector as shown in Figure 1a to share a common optical path. The laser beam is collimated, whereas the light from the LED is focused via a condenser lens to the pupil plane of the microscope objective (MO) in order to realize a Köhler illumination.

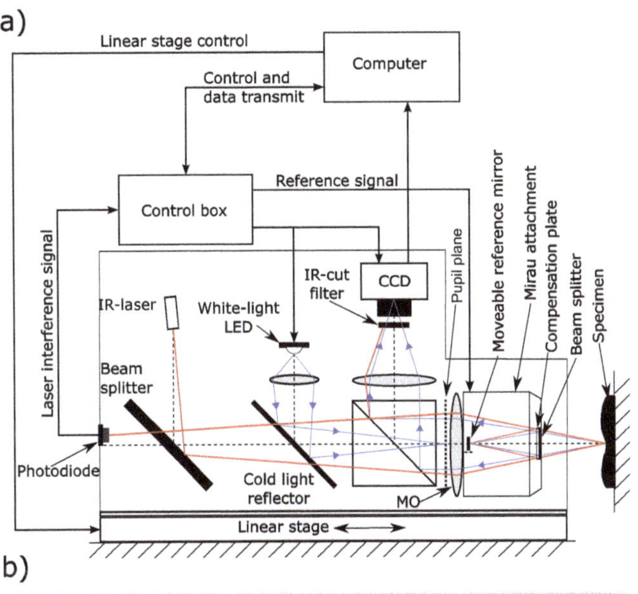

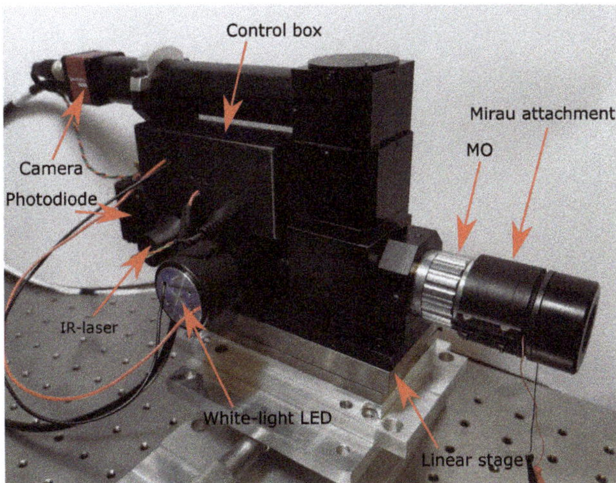

Figure 1. Schematic illustration (**a**) and photograph (**b**) of the Mirau CSI with a custom-built Mirau attachment comprising the oscillatable reference mirror in front of a commercial MO.

This leads to a collimated illumination of the surface, where each illuminated point in the pupil plane acts a secondary point source. The light reflected from the surface of the reference mirror as well as the scattered light from the sample are superimposed by a beam splitter plate and is then collected by the infinity corrected MO. The resulting interference pattern is then detected in the image plane by the CCD camera. The infrared light is separated from the temporally low coherent light by an IR-cut filter and recorded by a photodiode. Introducing a sinusoidal modulation of the optical path difference (OPD) by an oscillating reference mirror, the laser interferometer works as an IDS [14,16]. Both CSI and IDS work simultaneously and share the oscillating reference mirror. In order to reduce blurring of the captured camera frames, camera and LED triggering are synchronized with the mirror deflection, so that the camera frame is captured at the turning points of the oscillating reference mirror, where its motion speed is close to zero (visualized in the following Section by a trigger impulse shown in Figure 3). The control of the measurement process is provided by the control-box which is explained in more detail in [13,24]. As a consequence, a stroboscopic illumination is required to perform the measurement procedure. Short pulses of approximately 5–10 µs with overvoltage and overcurrent are generated by the LED to illuminate the surfaces of specimen and reference mirror with enough light. The IR-laser source is tilted relatively to the optical axis to be not blocked by the reference mirror located in the center of the optical axis. The interference signal obtained by the superimposed partial laser beams, which are reflected from the reference mirror and the surface to be measured, is detected by a photodiode. Only a small portion of the NA is used to focus and detect the laser spot by the MO. The practical realization of the sensor setup is depicted in Figure 1b.

As mentioned before, a MA is developed to realize a Mirau type setup with oscillating reference mirror for the IDS. Figure 2 shows a detailed photograph of the MA. It consists of three parts: the main body, an adjustment head and a clamping ring. In order to avoid dependencies on the manufacturing tolerances, the beam splitter plate is adjustably mounted to balance the interferometer arms. Therefore, the adjustment head including the plate beam splitter and the compensation plate is connected to the main body by a fine-thread, which has a pitch of 0.5 mm according to DIN 13-3.

Figure 2. Photograph of the custom-built MA with demounted adjustment head for better view of the used piezo actuator.

Further, a cover glass thickness compensated MO of Mitutoyo (G Plan Apo) with a magnification of 20× is used [25]. This MO provides a long working distance of 30.6 mm (while using a 3.5 mm thick cover glass) and an NA of 0.28. These properties enable an integration of the beam splitter plate and a compensation plate inside the MA avoiding any additional optical aberrations. In order to reduce dispersion, these two components have identical design parameters related to refractive index, geometrical dimensions, and surface quality. Therefore, these glasses are fabricated meeting high requirements. Both components are made of the same glass type (H-K9L) with a refractive index of $n_\mathrm{d} = 1.51680$ manufactured by CDGM [26]. The thickness tolerance is ± 2 µm, while the surface requirements are given by, e.g., flatness $\leq \lambda/10$, parallelism $< 30''$ and surface quality specified by a scratch and dig of 20-10. According to the MO specifications for the cover-glass, the thickness of both, the beam splitter plate and the compensation plate is 1.75 mm, respectively. To avoid an additional mount with inaccuracy due to manufacturing tolerances, these two elements are merged while the beam splitter coating is between the elements.

Respecting the mounting position of the reference mirror within the optical path, small dimensions of the mirror are needed to avoid obscurations. In addition, a mounting position as close as possible to the MO is chosen for the reference mirror to place it next to the pupil plane. As a result the shading is reduced as much as possible. The reference mirror is sinusoidally deflected by a miniature multilayer piezo actuator (PI PICMA PL0xx) [27]. With compact dimensions of $2\,\text{mm} \times 2\,\text{mm} \times 2\,\text{mm}$ this component yields a peak-to-peak amplitude of approximately $2.2\,\mu\text{m}$. A small piece of coated silicon wafer acting as flat reference mirror is placed on the piezo actuator (see Figure 2). The piezo actor deflects the reference mirror in axial direction with an oscillation frequency of $1\,\text{kHz}$, due to the need of a sufficient elongation for an error-free signal evaluation. Note that the oscillation frequency is limited by the used piezo controller (Thorlabs MDT694B).

The electrical connection wires of the piezo actuator are led out of the housing. Therefore, a bracket for the piezo is designed with two bars of the same width as the diameter of the wires to avoid additional obscuration. However, a high bending strength is necessary to suppress additional vibrations. The bracket is made of a glass bead reinforced polyamide (PA12-GB) and is manufactured by a selective laser sintering (SLS) process. This manufacturing method provides high accuracy in combination with low cost. The piezo generates a high waste heat during operation, so that these bars must be able to dissipate a high amount of heat. The operation at an excitation frequency of $1\,\text{kHz}$ is possible without any impact on the performance of the piezo actuator. However, a mount made of copper might be necessary for use with much higher excitation frequency in order to achieve a better heat management.

3. Signal Processing

The signal processing is done subsequent to the measurement procedure. The vibration compensating signal processing can be separated into two parts. In the first part, the correction of the distorted CSI signals occurs, whereby the evaluation of the IDS signal is performed to obtain the distance changes during the CSI measurement followed by the correction of the CSI datasets. These corrected CSI interference signals are then analyzed in the second step.

A two-beam interference signal can be described by

$$I_{\text{int}} = I_{\text{m}} + I_{\text{r}} + 2\sqrt{I_{\text{m}} I_{\text{r}}}|\gamma_{12}|\cos(\Phi), \tag{1}$$

where I_{m} and I_{r} represent the intensities of reflected beams from the measurement object and the reference mirror, respectively. $|\gamma_{12}|$ is the absolute value of the complex degree of coherence, which is <1 in case of the CSI depending on the use of temporally low coherent light. The phase Φ is represented by

$$\Phi_{\text{CSI}} = \frac{4\pi}{\lambda_{\text{LED}}}\Delta z_{\text{m}}(x,y) = \frac{4\pi}{\lambda_{\text{LED}}}(z_{\text{m}}(x,y) + z_{\text{ds}} + z_{\text{vib}}), \tag{2}$$

with the center wavelength of the illumination source λ_{LED} and the path length change Δz_{m} between the reference mirror and the surface caused by the surface structure $z_{\text{m}}(x,y)$, the depth scan z_{ds} and environmental vibrations z_{vib}. Assuming vibrations with oscillation frequencies below $1\,\text{kHz}$ these can be detected by the IDS. Its interference equation equals Equation (1) with $|\gamma_{12}| \approx 1$ and

$$\Phi_{\text{IDS}} = \frac{4\pi}{\lambda_{\text{LD}}}[\hat{z}_{\text{a}}\cos(2\pi f_{\text{a}} t) - \Delta z_{\text{m}}(x_0, y_0)], \tag{3}$$

where $\hat{z}_{\text{a}}$ and f_{a} represent, respectively, the amplitude and the frequency of the oscillating reference mirror, leading to an OPLM. x_0 and y_0 describe a certain point on the surface under investigation. Such an interference signal (blue curve) for an OPLM resulting from the electrical actuator signal $u(t)$ (green) and the related deflection of the reference mirror $z(t)$ (brown) is depicted in Figure 3. Assuming a linear deflection regarding the rising and falling flank (see Figure 3) of the sinusoidal reference mirror movement a constant fringe frequency $f_e = 4\pi \hat{z}_a f_a / \lambda_{LD}$ of the interference signal can be assumed. Therefore, the phase values of the corresponding sections of the interference signal are determined using a so-called single-point-DFT also known as lock-in method:

$$\Phi_{IDS} = \arg\left(\sum_{n=0}^{N_W-1} I(n)W(n)\exp\left(-i2\pi n \frac{f_e}{f_s}\right)\right),\tag{4}$$

with the window function $W(n)$ to reduce leakage, the number N_w of values covered by $W(n)$ and the sampling frequency f_s. Finally, the relationship between the phase and the distance value Δz_m is given by

$$\Delta z_m = \frac{\lambda_{LD}}{4\pi}\Delta\Phi_{IDS},\tag{5}$$

where $\Delta\Phi_{IDS}$ represents the difference of consecutive phase values determined according to Equation (4). This finally results in two height values per actor period [14,16].

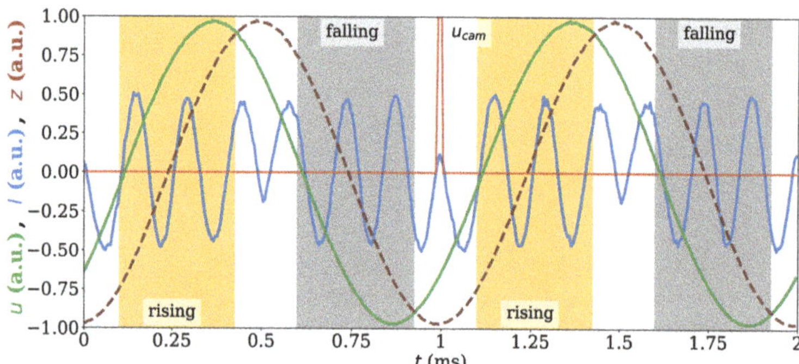

Figure 3. Measured IDS signal (blue) sampled during a CSI depth scan and the corresponding actuator excitation (brown). The excitation is caused by the voltage curve (green). The evaluation windows are marked by the colored areas where the nearly linear mirror deflection leads to a constant frequency f_e of the interference signal. u_{cam} represents the camera trigger signal at the turning point of the mirror deflection.

The correction of the CSI data occurs after the evaluation of the IDS signal is finished. For this purpose, the distance changes compared to the desired movement obtained from the IDS during a depth scan is used to rearrange the CSI data according to Figure 4. The re-arranging process leads to non-equidistant sampling points of the low coherent interference signals. However, in the processing of CSI signals by commonly utilized algorithms we expect an equidistant sampling of the correlograms. Therefore, an interpolation of the data is performed to obtain sample points at equidistant intervals. A trigonometric interpolation [28] provides an appropriate method as reported in [14]. The correction described above does not change the initial signal envelope. Consequently, the 3D topography can be obtained from the corrected correlograms by using widespread CSI signal processing algorithms which use the interference contrast and phase of the signal [29–32]. In this work, the CSI signals are evaluated by a phase evaluation algorithm using the lock-in technique similar to Equation (4) and additionally, an envelope detection using the Hilbert transform in order to increase the unambiguous range of two adjacent height values. In this way, an axial repeatability in the subnanometer range is achieved. These algorithms and their combination is described in more detail in [24].

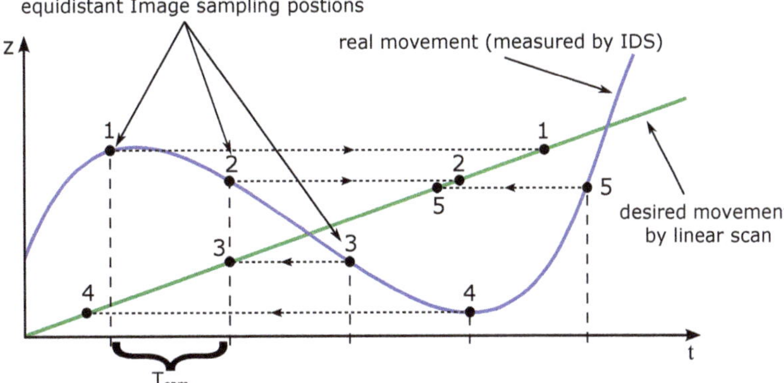

Figure 4. Schematic illustration of the correction algorithm. The CSI dataset is rearranged according to the real movement measured by the IDS (oriented to [24]).

4. Results and Discussion

Multiple measurements of different surface topographies are performed to validate the performance of the custom-built MA. In the following, the investigations are divided into two parts. At first, measurements are performed to validate the function of the CSI with a focus on the MA. These measurements are performed on a vibration damped table without applying external vibrations during the measurement. In addition, the measurement process in the second part is disturbed by an externally stimulated vibration and carried out without vibration damping. This part two deals with the capability to correct disturbances in CSI data sets.

4.1. Functioning of the Mirau Configuration

Different measurements are performed to illustrate the functioning of the MA. First, an aluminum mirror with a surface flatness less than $\lambda/10$ is used as a measuring object to investigate optical aberrations and systematic deviations. As mentioned before, a beam splitter plate and a compensation plate of high optical quality is used. Nevertheless, tilting of these optical elements with respect to the optical axis, e.g., due to installation tolerances that may occur during assembly, can cause dispersion. A possibility to determine dispersion effects is given by the investigation of the measured interference signals. Therefore, the envelopes of signals of the custom-built 20x Mirau objective are compared with those of a commercial Mirau objective (Nikon 10x CF IC EPI Plan DI). Both objectives are directly applied in the same optical setup to keep the results the comparable. The signal envelopes for both MOs are obtained from the aluminum mirror using a white as well as a red LED for illumination. As shown in Figure 5a, significant difference between the envelopes (red curves) of both MOs can be observed for white light illumination. While the course of the signals obtained by the commercial objective is symmetrical, the course corresponding to the MA is asymmetric. Furthermore, the interference signal is broadened compared to the signal of the commercial system, which is illustrated by a full width at half maximum (FWHM) of 2.54 µm for the MA and 1.52 µm for the Nikon Mirau objective. With reference to the investigations of de Groot and Colonna de Lega [33] and under the assumption that the Mitutoyo MO is corrected to a high degree and only minor optical aberrations are present, the broadening of the envelope probably is caused mainly by dispersion. This assumption is confirmed comparing Figures 5c,d, where the same measurements are performed but using a red LED. Here, as a consequence of a narrower bandwidth the envelopes for both MOs are broadened compared to the signals obtained for a white LED. However, no significant difference is visible between the interference signals of both MOs. This is also corroborated by the difference of FWHMs obtained from the respective envelopes of the interference signals, which decrease from 1.02 μm for the white LED to 0.02 μm using the red LED as illumination source.

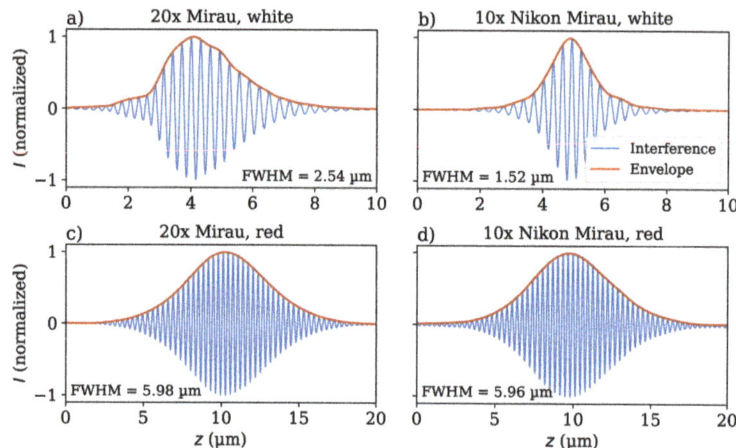

Figure 5. Interference signals obtained from an aluminum mirror using the custom-built (**a,c**) as well as the commercial Mirau objective (**b,d**).

The measurement of the aluminum mirror shows systematic deviations in the measured surface topography, which still remain when the measuring field is laterally shifted. In order to separate the systematic deviations from the topography of the measured object, nine measurements are performed on latterly shifted locations at the aluminum mirror used as measuring object. The pixel-wise averaging of the nine measured surface topographies leads to the systematic deviation of each pixel, similar to the procedure used in [34–36]. This deviation is also known as residual flatness and includes the systematic deviations due to the unevenness of the reference mirror. In Figure 6a, the illustrated residual flatness is measured using a white LED, whereas for the result depicted in Figure 6b, a red LED is used. Similarities in both topographies are apparent immediately. The topography in Figure 6a is slightly higher compared to that depicted in Figure 6b. This is also illustrated by calculating the standard deviations $\sigma_w = 3.47$ nm for the surface obtained using the white LED and the red one with $\sigma_r = 3.25$ nm. However, the specification of the standard deviation to represent the residual flatness is misleading, since the determined systematic deviations differ stronger at the edges compared to the center (see Figure 6). The standard deviation is reduced to 0.995 nm for a subtraction of the two topographies from each other and a predominantly uniform shape occurs. Consequently, deviations in flatness of the Mirau CSI are not mainly caused by dispersion and can be attributed to other influences such as flatness deviations or contaminations of the reference mirror or other system-related wavefront errors. The residual flatness can be used for the calibration of the system by pixel-wise subtracting the residual flatness from each performed topography measurement. An application is shown in Figure 6c, where an additional measurement of the aluminum mirror is calibrated by subtracting the residual flatness shown in Figure 6a from the measured topography. As depicted in the Figure 6c a nearly uniform topography occurs with a standard deviation of $\sigma_{corr} = 0.62$ nm. Consequently, subtracting the residual flatness is a suitable approach for correcting the measured topography data. Therefore, this correction approach is considered in all following results in order to reduce systematic errors of the measured surface topographies.

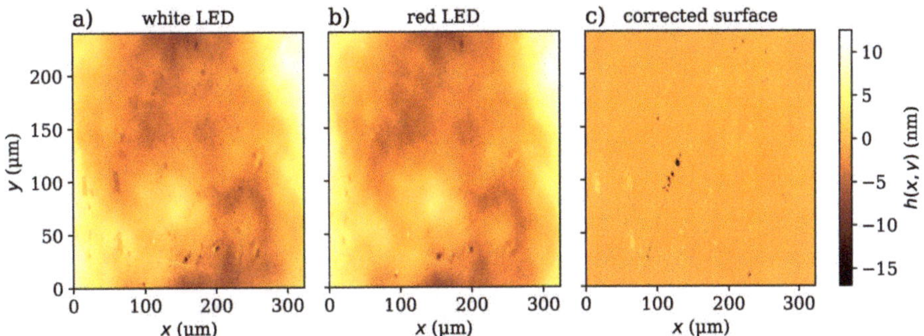

Figure 6. Residual flatness obtained by averaging nine slightly shifted measured topography data sets (**a**) acquired with a white and (**b**) with red light illumination and (**c**) corrected topography of an additionally measured mirror surface by subtracting the residual flatness.

In addition, a resolution standard (type RS-M from Simetrics GmbH) comprising several gratings of period lengths from 4 µm to 800 µm with a nominal step height of 90 nm is used [37]. The results are shown in Figure 7.

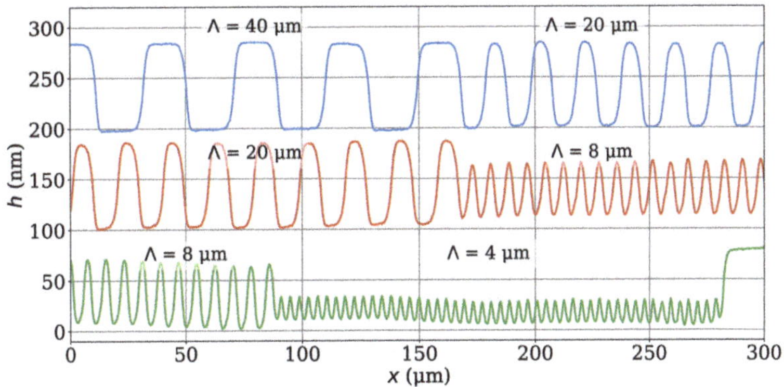

Figure 7. Profiles of RS-M standard obtained by 20x Mirau with different grating periods Λ of 40 µm and 20 µm (blue curve), 20 µm and 8 µm (red curve) and 8 µm and 4 µm (green curve).

The gratings with a period length of 40 µm and 20 µm (blue profile) are well resolved with minor measurement deviations apart from rounded edges. The step height determined following the procedure described in DIN EN ISO 5436-1 is 85.9 nm for the 40 µm and 81.5 nm for the 20 µm structure. In the red profile an asymmetric constriction is observable for the grating of $\Lambda = 8$ µm leading to a measured step height of 50.3 nm. This asymmetry occurs due to a sharp-combed structure, where the width of the upper plateaus are smaller than those of the lower plateaus [37]. Consequently, a low pass filtering due to the NA of 0.28 for the MA setup has a stronger effect on the upper plateaus compared to the lower ones. According to the Rayleigh criterion, the lateral resolution of the Mirau objective is 1.36 µm for an evaluation wavelength of 625 nm and an NA of 0.28. Therefore, the smallest grid with a period length of 4 µm (green profile) is still well resolved as expected. However, the step height for the 4 µm grating is clearly decreased to 18.3 nm. Obviously, the low-pass filtering effect is strongest for the 4 µm structure. In addition, an asymmetric constriction similar to the grating of $\Lambda = 8$ µm appears, which is again a result of a sharp-combed structure.

In an industrial context, rough surfaces often are encountered in topography measurement applications. For this reason, roughness standards are frequently used for the realistic characterization of a measuring system. Accordingly, the system is also characterized using the superfine roughness standard KNT4070/03 by Halle GmbH [38]. According to DIN EN ISO 4288, the roughness parameters R_a, R_q, and R_z are convenient to characterize surface roughness. However, these parameters are specified for a profile length of 1.25 mm, which is significantly longer than the maximum edge length of the measurement field of 324 µm. Therefore, these parameters are not used in this work to characterize the roughness of the measured profile. Taking into account, that reference values for such roughness standards are usually obtained by tactile stylus instruments, the profile measured by the custom-built 20x Mirau CSI is compared to the same profile section sampled by the tactile stylus instrument Marsurf GD26 from Mahr, which is integrated in a multisensor measuring system [39]. As shown in Figure 8, the profile resulting from the CSI measurment agrees quite well to the tactile measurment result.

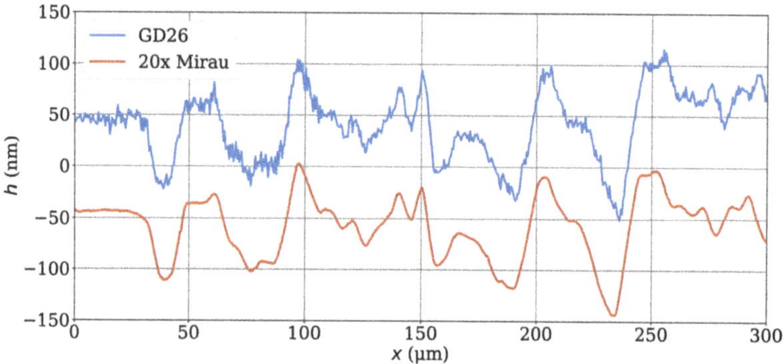

Figure 8. Profiles measured on the superfine roughness standard KNT4070/03. The blue profile is obtained by the tactile stylus instrument GD26 and the red one below by the custom-built 20x Mirau CSI.

However, a low-pass filtering of the profile obtained by the 20x Mirau occurs due to the relatively low NA of 0.28, resulting in slight deviations from the tactile reference profile.

4.2. Validation of Vibration Compensation Capabilities of the MA

In the following, a measurement of a sine standard with 100 µm period length and 1 µm peak-to-valley amplitude is performed to illustrate the vibration compensation capabilities of the Mirau setup. For this purpose, two measurements are sequentially executed. First, the sinusoidal surface topography is measured in a disturbance-free environment. The result is depicted in Figure 9a.

Then, a well-defined vibration is applied. For this purpose, the object to be measured is sinusoidally deflected by a piezo-driven stage with an amplitude of 200 nm and an oscillation frequency of 8 Hz. Figure 9b shows the impact of the disturbance on the measured surface topography. There is a significant difference to the disturbance-free surface topography, and the sinusoidal structure can only be suspected. Besides, the CSI signal the laser interferometric signal of the IDS is detected during the depth scanning process, leading to the course of height values depicted in Figure 10. Due to the acquisition rate of 1000 height values per second, the changing height caused by the depth scan (linear part of the course) and the superimposed oscillation of the vibration could be detected by the IDS. Consequently, the frames captured by the CCD camera can be resorted according to the distance values measured by the IDS. The evaluation of the sorted and trigonometrically interpolated CSI signals using a phase analyzing algorithm leads to the surface topography depicted in Figure 9c. A comparison with the unaffected surface topography (see Figure 9a) shows hardly any deviation. The determined peak-to-valley amplitude of the surface obtained from the corrected data corresponds to 0.998 µm, while the measured amplitude of the surface topography measured without disturbance is 0.992 µm. In both measurements the period length results in 100 µm. Therefore, the influences of the disturbance on the measured surface is successfully corrected.

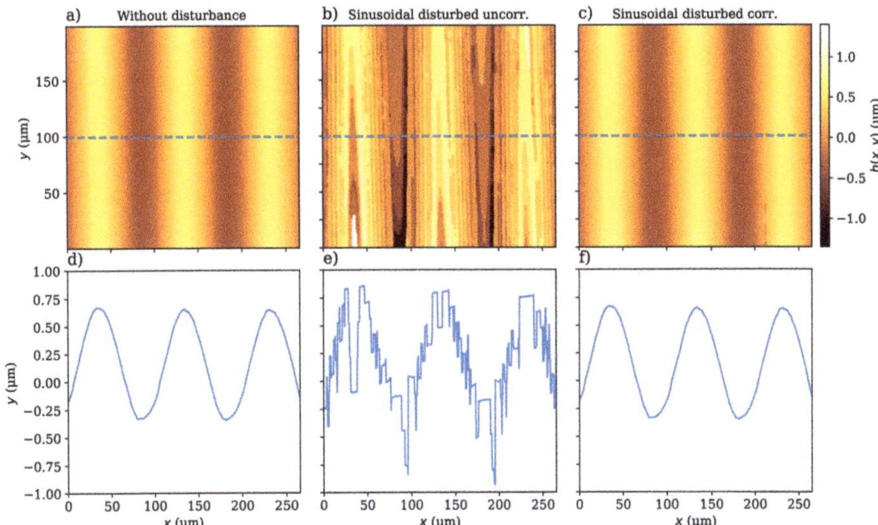

Figure 9. Measured sinusoidal standard with 1 µm peak-to-valley amplitude and 100 µm period length. (**a**) Topographies obtained without disturbances on a vibration-damping table, (**b**) disturbed by a sinusoidal vibration with an amplitude of 200 nm and an oscillation frequency of 8 Hz, (**c**) same topography as shown in panel (**b**) but corrected according to the described compensation method, (**d**–**f**) the corresponding profiles marked by the blue dashed line in the (**a**–**c**) surface topographies.

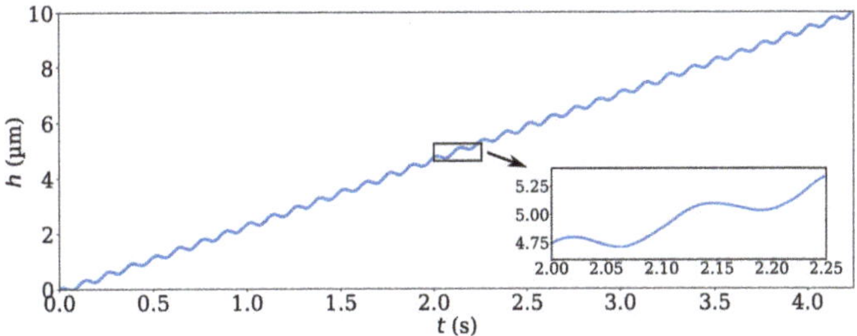

Figure 10. Measured IDS signal of a depth scan of 10 µm disturbed by a sinusoidal oscillation.

5. Conclusions and Outlook

In this contribution, a CSI comprising a Mirau objective is introduced, where a custom-built interference device is attached to a commercial microscope objective. An oscillating reference mirror inside the Mirau objective's attachment enables the IDS measuring process in parallel to the conventional CSI depth scan. Thus, the changed distance between CSI and surface under investigation including influences caused by external vibrations are measurable by the IDS. This enables a correction of the disturbance as shown for a sinusoidal standard. Due to its advantages, the Mirau setup is the preferred and most frequently used interferometer type in industry. Thus, to the best of the authors knowledge, this is a novel combination, enabling the industrial application of the previously developed method for vibration compensation. Furthermore, the proper optical functioning of the custom-built Mirau objective is investigated by comparative measurements on various surface structures using different LEDs and reference instrumentation such as a tactile stylus instrument and a 10x Mirau objective lens from Nikon Inc. The measured surface structure of the superfine roughness standard and the gratings of the RSM standard are well resolved as expected for a CSI with an NA of 0.28. A comparison between interference signals obtained by the custom-built Mirau objective using different LEDs of different spectral bandwidth reveals that interference signals obtained for a white LED illumination suffer from dispersion. Moreover, low systematic deviations in the range of ± 10 nm occur, which are basically independent from the dispersion effect. Since these systematic deviations are reproducible, it can be eliminated from the measured surface structure after calibration. These deviations and disturbances might be a result of misalignment in the custom-built objective. In further investigations, the alignment of the optical elements will be measured in a custom-built interferometric setup.

In this work, a disturbance affecting the interference signals of each camera pixel in the same way is assumed. However, the influence of external vibrations might be different across the surface to be measured. Therefore, multiple IDS can be applied to measure vibrations on laterally shifted locations on the surface under investigation. This is part of future developments. The laser beam of the IDS is currently tilted to be able to pass the reference mirror of the Mirau objective. Consequently, a decreased intensity of the reflected laser light occurs with increasing surface slope. This can probably be optimized using critical illumination for the IDS.

Author Contributions: Conceptualization, H.S.; investigation, H.S. methodology, H.S.; software, P.G.; validation, H.S., P.G. and S.H.; formal analysis, H.S. and S.H. writing—original draft preparation, H.S.; writing—review and editing, H.S., P.G., S.H. and P.L.; supervision, P.L.; funding acquisition, P.L. All authors have read and agreed to the published version of the manuscript.

Funding: The support of this project under the WIPANO program (grant number 03THW10K24) by the German Ministry of Economy and Energy is gratefully acknowledged.

Institutional Review Board Statement: Not applicable.

Informed Consent Statement: Not applicable.

Data Availability Statement: The data presented in this study are available on request from the corresponding author.

Conflicts of Interest: The authors declare no conflicts of interest. The funders had no role in the design of the study; in the collection, analyses, or interpretation of data; in the writing of the manuscript; and in the decision to publish the results.

References

1. Kino, G.S.; Corle, T.R. *Confocal Scanning Optical Microscopy and Related Imaging Systems*; Academic Press: Cambridge, MA, USA, 1996.
2. Lehmann, P.; Tereschenko, S.; Xie, W. Fundamental aspects of resolution and precision in vertical scanning white-light interferometry. *Surf. Topogr. Metrol. Prop.* **2016**, *4*, 024004. [CrossRef]
3. De Groot, P. Principles of interference microscopy for the measurement of surface topography. *Adv. Opt. Photonics* **2015**, *7*, 1–65. [CrossRef]
4. Schake, M.; Lehmann, P. Quadrature-based interferometry using pulsed RGB illumination. *Opt. Express* **2019**, *27*, 16329–16343. [CrossRef]
5. Gollor, P.; Schake, M.; Tereschenko, S.; Roetmann, K.; Mann, K.; Schäfer, B.; Uhlrich, G.; Haberland, M.; Lehmann, P. Kombination eines neuartigen Doppelpuls-RGB-Interferometers mit einem Hartmann-Shack-Wellenfrontsensor zur dynamischen flächenhaften Topographieerfassung. *Tm-Tech. Mess.* **2020**, *87*, 523–534. [CrossRef]
6. Kühn, J.; Colomb, T.; Montfort, F.; Charrière, F.; Emery, Y.; Cuche, E.; Marquet, P.; Depeursinge, C. Real-time dual-wavelength digital holographic microscopy with a single hologram acquisition. *Opt. Express* **2007**, *15*, 7231–7242. [CrossRef] [PubMed]
7. Wu, D.; Zhu, R.H.; Chen, L.; Li, J.Y. Transverse spatial phase-shifting method used in vibration-compensated interferometer. *Optik* **2004**, *115*, 343–346. [CrossRef]
8. Jiang, X.; Wang, K.; Gao, F.; Muhamedsalih, H. Fast surface measurement using wavelength scanning interferometry with compensation of environmental noise. *Appl. Opt.* **2010**, *49*, 2903–2909. [CrossRef] [PubMed]

9.   Schäfer, P.; Broschart, D.; Seewig, J. Aktive Schwingungsdämpfung eines Weißlichtinterferometers. *Tm-Tech. Mess.* **2013**, *80*, 16–20. [CrossRef]

10. Teale, C.; Barbastathis, G.; Schmidt, M.A. Vibration compensated, scanning white light interferometer for in situ depth measurements in a deep reactive ion etcher. *J. Microelectromech. Syst.* **2019**, *28*, 441–446. [CrossRef]

11. Deck, L.L. Suppressing phase errors from vibration in phase-shifting interferometry. *Appl. Opt.* **2009**, *48*, 3948–3960. [CrossRef]

12. Kiselev, I.; Kiselev, E.I.; Drexel, M.; Hauptmannl, M. Noise robustness of interferometric surface topography evaluation methods. Correlogram correlation. *Surf. Topogr. Metrol. Prop.* **2017**, *5*, 045008. [CrossRef]

13. Tereschenko, S.; Lehmann, P.; Gollor, P.; Kuehnhold, P. Robust vertical scanning white-light interferometry in close-to-machine applications. In *Optical Measurement Systems for Industrial Inspection IX*; International Society for Optics and Photonics: Bellingham, WA, USA, 2015; Volume 9525, p. 95250Q.

14. Tereschenko, S.; Lehmann, P.; Zellmer, L.; Brueckner-Foit, A. Passive vibration compensation in scanning white-light interferometry. *Appl. Opt.* **2016**, *55*, 6172–6182. [CrossRef] [PubMed]

15. Tereschenko, S.; Lehmann, P.; Gollor, P.; Kuehnhold, P. Vibration compensated high-resolution scanning white-light Linnik-interferometer. In *Optical Measurement Systems for Industrial Inspection X*; International Society for Optics and Photonics: Bellingham, WA, USA, 2017; Volume 10329, p. 1032940

16. Hagemeier, S.; Tereschenko, S.; Lehmann, P. High-speed laser interferometric distance sensor with reference mirror oscillating at ultrasonic frequencies. *Tm-Tech. Mess.* **2019**, *86*, 164–174. [CrossRef]

17. Sasaki, O.; Okazaki, H. Sinusoidal phase modulating interferometry for surface profile measurement. *Appl. Opt.* **1986**, *25*, 3137–3140. [CrossRef]

18. Zhang, Q.; Pan, W.; Cross, L.E. Laser interferometer for the study of piezoelectric and electrostrictive strains. *J. Appl. Phys.* **1988**, *63*, 2492–2496. [CrossRef]

19. Martini, G. Analysis of a single-mode optical fibre piezoceramic phase modulator. *Opt. Quantum Electron.* **1987**, *19*, 179–190. [CrossRef]

20. De Groot, P. Design of error-compensating algorithms for sinusoidal phase shifting interferometry. *Appl. Opt.* **2009**, *48*, 6788–6796. [CrossRef]

21. Schulz, M.; Lehmann, P. Measurement of distance changes using a fibre-coupled common-path interferometer with mechanical path length modulation. *Meas. Sci. Technol.* **2013**, *24*, 065202. [CrossRef]

22. Knell, H.; Laubach, S.; Ehret, G.; Lehmann, P. Continuous measurement of optical surfaces using a line-scan interferometer with sinusoidal path length modulation. *Opt. Express* **2014**, *22*, 29787–29798. [CrossRef]

23. Kino, G.S.; Chim, S.S. Mirau correlation microscope. *Appl. Opt.* **1990**, *29*, 3775–3783. [CrossRef] [PubMed]

24. Tereschenko, S. Digitale Analyse Periodischer und Transienter Messsignale Anhand von Beispielen aus der Optischen Präzisions-messtechnik. Ph.D. Thesis, University of Kassel, Kassel, Germany, 2018

25. Mitutoyo Corporation. Microscope Units and Objectives (UV, NUV, Visible and NIR Region). 2019. Available online: https://www.mitutoyo.com/wp-content/uploads/2020/12/E14020.pdf (accessed on 11 October 2021).

26. CDGM Glass Company Ltd. Datasheet H-K9L. 2021. Available online: http://cdgmglass.com/Portals/0/CDGMSearch2/pdf/H-K9L.pdf (accessed on 11 October 2021).

27. PI Ceramic GmbH. Datasheet PICMA Chip Actuator PL0xx. 2017. Available online: https://static.piceramic.com/fileadmin/user_upload/physik_instrumente/files/datasheets/PL0xx_Datasheet.pdf (accessed on 11 October 2021).

28. Reichel, L.; Ammar, G.; Gragg, W. Discrete least squares approximation by trigonometric polynomials. *Math. Comput.* **1991**, *57*, 273–289. [CrossRef]

29. De Groot, P.; Deck, L. Surface profiling by analysis of white-light interferograms in the spatial frequency domain. *J. Mod. Opt.* **1995**, *42*, 389–401. [CrossRef]

30. Larkin, K.G. Efficient nonlinear algorithm for envelope detection in white light interferometry. *JOSA A* **1996**, *13*, 832–843. [CrossRef]

31. Harasaki, A.; Schmit, J.; Wyant, J.C. Improved vertical-scanning interferometry. *Appl. Opt.* **2000**, *39*, 2107–2115. [CrossRef] [PubMed]

32. Fleischer, M.; Windecker, R.; Tiziani, H.J. Fast algorithms for data reduction in modern optical three-dimensional profile measurement systems with MMX technology. *Appl. Opt.* **2000**, *39*, 1290–1297. [CrossRef] [PubMed]

33. De Groot, P.; Colonna de Lega, X. Signal modeling for low-coherence height-scanning interference microscopy. *Appl. Opt.* **2004**, *43*, 4821–4830. [CrossRef] [PubMed]

34. Hagemeier, S.; Lehmann, P. High resolution topography sensors in a multisenor measuring setup. In *Optical Measurement Systems for Industrial Inspection XI*; International Society for Optics and Photonics: Bellingham, WA, USA, 2019; Volume 11056, p. 110563I.

35. De Groot, P.; DiSciacca, J. Surface-height measurement noise in interference microscopy. In *Interferometry XIX*; International Society for Optics and Photonics: Bellingham, WA, USA, 2018; Volume 10749, p. 107490Q.

36. Giusca, C.L.; Leach, R.K.; Helary, F.; Gutauskas, T.; Nimishakavi, L. Calibration of the scales of areal surface topography-measuring instruments: Part 1. Measurement noise and residual flatness. *Meas. Sci. Technol.* **2012**, *23*, 035008. [CrossRef]

37. Krüger-Sehm, R.; Frühauf, J.; Dziomba, T. Determination of the short wavelength cutoff of interferential and confocal microscopes. *Wear* **2008**, *264*, 439–443. [CrossRef]

38. Halle GmbH. Calibration Standards for Contact Stylus Instruments, Line of Products KNT 4070/03. 2009. Available online: http://www.halle-normale.de/pdf/Prospektseiten/englisch/17%20Ps-KNT4070_03_Bi_6-6_GB.pdf (accessed on 11 October 2021).
39. Hagemeier, S.; Schake, M.; Lehmann, P. Sensor characterization by comparative measurements using a multi-sensor measuring system. *J. Sens. Sens. Syst.* **2019**, *8*, 111–121. [CrossRef]

Article

Dynamic, Adaptive Inline Process Monitoring for Laser Material Processing by Means of Low Coherence Interferometry

Fabian Zechel [1,*], Julia Jasovski [1] and Robert H. Schmitt [1,2]

[1] Fraunhofer Institute for Production Technology IPT, Steinbachstr. 17, 52074 Aachen, Germany; julia.jasovski@ipt.fraunhofer.de (J.J.); R.Schmitt@wzl.rwth-aachen.de (R.H.S.)

[2] WZL I RWTH Aachen University, Campus-Boulevard 30, 52074 Aachen, Germany

[*] Correspondence: fabian.zechel@ipt.fraunhofer.de; Tel.: +49-241-8904-543

Abstract: Surface laser structuring of electrical steel sheets can be used to manipulate their magnetic properties, such as energy losses and contribute to a more efficient use. This requires a technology such as low coherence interferometry, which makes it possible to be coupled directly into the existing beam path of the process laser and enables the possibility for an 100% inspection during the process. It opens the possibility of measuring directly in the machine, without removing the workpiece, as well as during the machining process. One of the biggest challenges in integrating an LCI measurement system into an existing machine is the need to use a different wavelength than the one for which the optical components were designed. This results in an offset between the measurement and processing spot. By integrating an additional scanning system exclusively for the measuring beam and developing a compensation model for the non-linear spot offset, this can be adaptively corrected by up to 98.9% so that the ablation point can be measured. The simulation model can also be easily applied to other systems with different components and at the same time allows further options for in-line quality assurance.

Keywords: electrical steel; optical coherence tomography; OCT; scanning; process monitoring; laser material processing; spot compensation; low coherence interferometry; LCI

Citation: Zechel, F.; Jasovski, J.; Schmitt, R.H. Dynamic, Adaptive Inline Process Monitoring for Laser Material Processing by Means of Low Coherence Interferometry. *Appl. Sci.* **2021**, *11*, 7556. https://doi.org/10.3390/app11167556

Academic Editor: Andreas Fischer

Received: 19 July 2021
Accepted: 16 August 2021
Published: 18 August 2021

Publisher's Note: MDPI stays neutral with regard to jurisdictional claims in published maps and institutional affiliations.

Copyright: © 2021 by the authors. Licensee MDPI, Basel, Switzerland. This article is an open access article distributed under the terms and conditions of the Creative Commons Attribution (CC BY) license (https://creativecommons.org/licenses/by/4.0/).

1. Introduction

Saving energy is becoming increasingly important. Especially in the area of generation and transformation of electrical energy, a considerable loss is to be expected. In the case of classic electrical sheet, several layers of steel and insulation materials are stacked. Electrical steel is used in a wide range of applications, from transformers and sensors to motors and generators [1–4]. One of the most important aspects of electric steel is the energy density. Unfortunately, a high energy density also leads to an increased energy loss. At the same time, noise generation is also increased due to, among other things, long-term delamination of the steel stack [5–7].

One way to improve these factors is to laser-texture the steel sheets. Laser micro structuring is an excellent solution due to its capacity to create functional surface structures on a micrometer scale [8–15]. The aim of the Horizon 2020 research project ESSIAL is, among other things, to improve the magnetic properties, as described above, by laser structuring of the steel sheets. The structures being considered are on grain-oriented (GO) electrical steel with the spatially width in the range of 25–500 μm and 1–25 μm depth range, as well as structures on grain non-oriented (GNO) in the range of 10–25 μm in width and 1–50 μm in depth [16–18].

To ensure that the target parameters of the process are met, a monitoring system is required that also minimizes the influence on the process and the machine itself and can be used to constantly observe the process and to avoid production rejects and to make production even more resource-efficient [19,20]. The low coherence interferometry (LCI) allows here the measurement of the ablation and thus the over surface structures

---

without the need to remove the workpiece from the machine and at the same time uses the same beam path as the existing laser structuring process. LCI is an extension of classical optical coherence tomography (OCT) [21]. OCT was developed for ophthalmology in the early 1990s and has already been used in technical applications for several years [22–26]. While OCT provides high resolution imaging results for tomographic images, using the LCI method allows the measurement of the surface distance and thus the derivation of the topology of the surface on the workpiece [27,28]. The optical components of a laser machine are designed for a specific wavelength range, and the wavelength of the process laser is located in this range. The energy of the process laser is capable of destroying the components of the measuring system, so that the two systems must be separated in a certain way. A beam splitter can be used to achieve this, but the measurement wavelength must still be different from the wavelength of the process laser. This results in a non-linear offset between the spot positions of the two beams, so that simultaneous measurement and processing of the same position is no longer possible. The distance between the two spots increases as the beam moves further away from the center of the objective lens. In principle, this can be compensated by a color-corrected objective lens, but not completely or the choice of these objectives are very limited [29,30]. For this to be done, it is necessary to develop a method that allows this non-linear offset to be compensated adaptively during the process, thus allowing direct measurement of the material ablation. It must also be investigated how such compensation can be implemented and controlled.

The strategy and the development of a system that meets these requirements is described in the following sections. The goal of the ESSIAL research project is to integrate a measuring system into the laser process. For this purpose, a laboratory system will be developed and then scaled up step by step to complete machine integration. In this paper the focus is on the compensation of the spot offset caused by the different wavelengths between the processing laser and the measuring system. In this way, a measuring system is developed that enables 100% quality control while enabling use on an industrial scale.

2. Measuring Principle

The measuring principle of the LCI is based on the interferometric evaluation of the change in optical path length. One big advantage of the LCI method is the independence between axial resolution σ_z and lateral resolution $\sigma_{(x,y)}$. While the axial resolution and the axial measuring range depends on the used light source, shown in Equations (1) and (2), the used scanning lens and scanning system have a big impact on the lateral resolution, shown in Equation (3) [22].

Assuming a Gaussian beam axial resolution and range and lateral resolution are defined by:

$$\sigma_z = \frac{2 \ln 2}{\pi} \frac{\lambda_0^2}{\Delta\lambda}, \tag{1}$$

$$\Delta_z = \frac{N}{4n} \frac{\lambda_0^2}{\Delta\lambda}, \tag{2}$$

$$\sigma_{x,y} = \frac{\Delta\lambda}{\pi} \frac{f}{d}, \tag{3}$$

where $\Delta\lambda$ is the spectral width and λ_0 the central wavelength of the light source, n the average refractive index of the sample, N is the number of used pixels of the detector, d is the beam diameter at objective lens entry and f is the focal length of the used objective lens [29,31]. A nomenclature of the dimensions and quantities used can be found in the Appendix A.

3. Development of a Laboratory Setup

The challenges of integrating a measurement system into an existing beam path are complex. Different wavelengths must be used so that a separation can be made between the measuring system and the process laser. At the same time, the optics used are designed only for the process laser, so that strong chromatic aberrations must be expected here [28]. Thus, a suitable choice of the measuring wavelength is no longer trivial. An ultra-short pulse laser (USP) with a central wavelength of 1030 nm is used for the machining process. The laser has a pulse width of 500 fs and a pulse frequency of 10 kHz.

Based on the parameters of the laser, the scanner and the optics used, the central wavelength of the LCI measurement system can now be determined. For this purpose, different potential wavelengths were first tested by means of an optical simulation with the replica of the machine's beam path. At the same time, one of the most important components is the beam splitter, which is intended to separate the two beams. Since the beam path is optimized for a wavelength of 1030 nm, the wavelength of the measuring system should be as close to this as possible. As shown in Equation (2), a high spectral width of the light source is advantageous. To take these boundary conditions into account, a beam splitter with a steep slope is necessary.

The combination of the NFD01–1040 beam splitter from Semrock and the Superlum SLD-471UBB was identified as the optimum combination, see Figure 1. In the use case considered here, the smallest structure is at a depth of 1 µm, so that the resolution of the measuring system must cover this. Equation (1) shows a theoretical axial resolution of 3.632 µm for the light source used with a central wavelength of 930 nm and a FWHM of 115 nm. It indicates the ability of the system to differentiate between different layers tomographically, but here a topographic measurement is made. For topographic evaluations, the repeatability must be taken into account, which is 0.36 µm, as will be shown later. Therefore, all structures of the use case can be resolved. To ensure optimal utilization of the broadband light, a suitable spectrometer is required that can capture the entire spectral width.

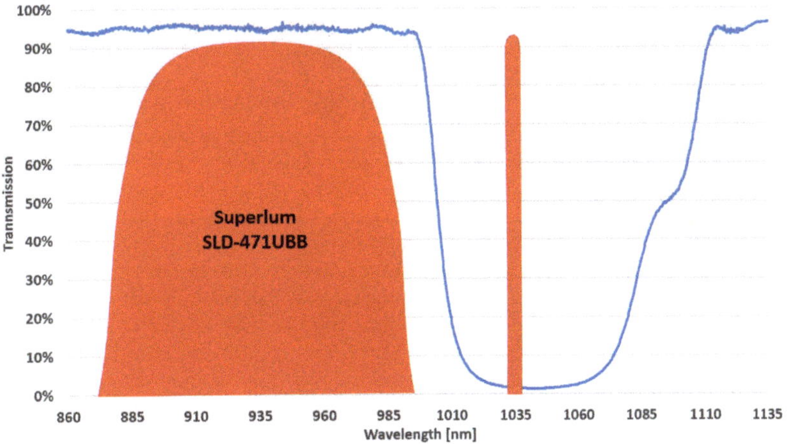

Figure 1. Transmission curve of the beam splitter with drawn-in spectra of the process laser and the selected light source.

3.1. Development of the Spectrometer

The spectrometer is one of the central components of the SD-LCI. In the spectrometer, the superimposed signals from the measuring and reference arms are broken down into their spectral components and evaluated. It consists of an optical grating to diffract the radiation, a lens package to focus the diffracted spectrum and the camera on whose line the focus is set. Simulations were used to determine the lens package in the spectrometer. The aim of the simulation is to image the measurement spectrum on the camera line with the smallest possible spot sizes, taking into account all pixels of the camera as well as the entire wavelength range of the light source.

To avoid a loss of intensity, the beam diameter used to illuminate the grating was chosen to be 10 mm. This ensures that the light does not come into contact with the non-optical components of the spectrometer. Initially, some lens configurations were simulated with an automated simulation; for cost reasons, only standard lenses were used.

The three best configurations were then further optimized manually. The parameters considered here were the size of the Airy-disc, the diffraction limit and the RMS of the spot radii. In a further step, the behaviour of the systems was compared with the MTF to achieve an optimal contrast ratio. Figure 2 shows a schematic sketch of the optical components of the optimized system.

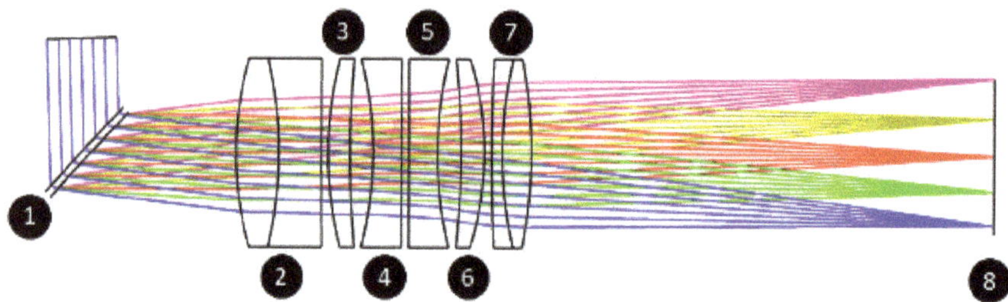

Figure 2. Final system model of the spectrometer beam shaping. The installed components are as follows: transmission grating NIR-1500-903 from Ibsen Photonics (1); lens package from Thorlabs consisting of AC254-125-B, LE1104, LC1120, LC1120, LE1156 and AC254-100-B (2 to 7); line scan camera AViiVA M2 CL 2010 from e2v (8).

Based on the simulation with Zemax, a system for beam shaping within the spectrometer could be identified and already optically designed. Thereby the boundary conditions for the mechanical construction of the spectrometer are determined, so that a following construction can be performed.

3.2. Offline Laboratory Setup and Characterisation

The integration process of an LCI measuring system into a laser machine is not trivial, in order to be able to make adjustments at an early stage, if there are deviations in the desired parameters, it is advisable to first build a laboratory system.

A setup based on a Thorlabs LSM03-BB lens was constructed. Although the optical path is significantly shorter than later in the system, this setup can be used to characterize the light source used and the spectrometer. These two components will later be operated unchanged with the target components. The final system uses the same spectrometer and the same light source, but a lens developed for laser material processing is used.

In order to perform a system characterization, it must be ensured that the intensity of light arriving at the line camera in the spectrometer does not lead to an overexposure of this sensor, see Figure 3. This overexposure would make the obtained results unsuitable. At the same time, throttling the output power of the light source leads to a changed spectrum and to a change in the optical properties of the system.

To avoid this, a Thorlabs V800A fiber optic attenuator (FOA) was installed directly after the light source, which allows the output power to be attenuated without changing the spectrum of the light. An overload of the sensor in the spectrometer, as shown in the left picture in Figure 3, leads to the fact that the Fourier transformation cannot determine the frequency spectrum correctly, ghost frequencies are present, shown in Figure 3 on the right side. The result is not usable, because no utilizable information is available. By throttling the FOA, the amount of light is reduced so that the full dynamic range of the spectrometer can be used and thus the spectrum is correctly resolved by the Fourier transform. The degree of this throttling depends on the material to be measured. Since this provides a usable result in the Fourier transformation, the peak in the spectrum generated by the surface can now be identified and later translated into a valid distance information. This ensures that the peak produced by the surface can be detected clearly and correctly later on, see Figure 4.

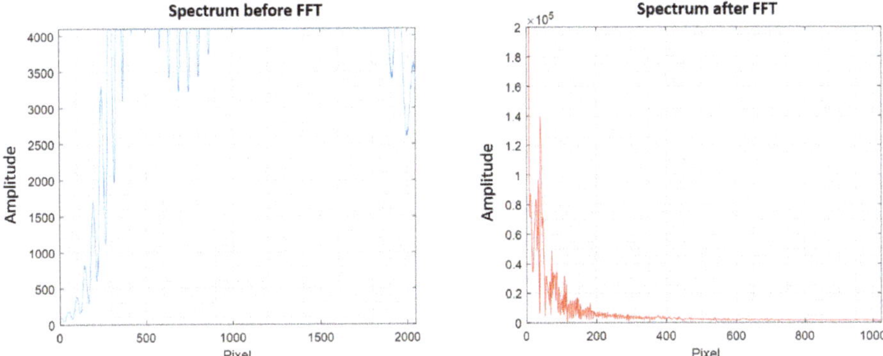

Figure 3. Measurement of a plane mirror without attenuation of the light intensity. On the left side the raw spectrum is found, a large part of the detector line is overexposed. On the right side the artifact sampled Fourier transformation of the signal is shown.

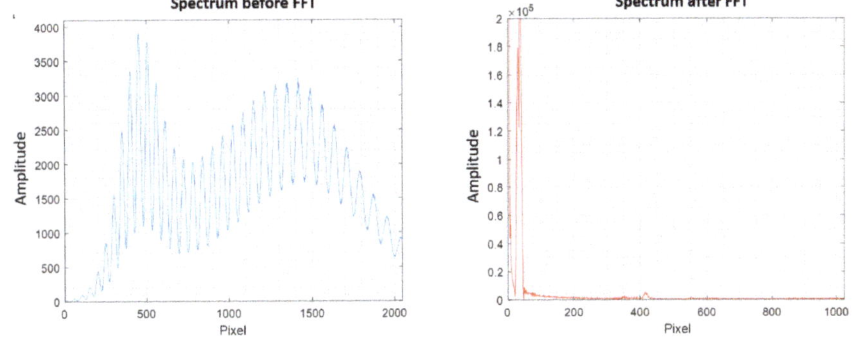

Figure 4. Measurement of a plane mirror with adjusted attenuation of the light intensity. On the left side the raw spectrum is shown. On the right side the peak of the surface is clearly visible.

A plane mirror is used to determine the measuring range of the system. This mirror is located in the focal point of the objective and on a piezo stepper, which can move the mirror in vertical direction. The piezo stepper (Physik Instrumente (PI) N-381.3A) used here has a maximum travel of 30 mm, with a resolution of 20 nm. This means that both the travel and the resolution are at least one order of magnitude better than the system can theoretically perform.

The mirror was moved by the piezo stepper with a step size of 0.1 mm and measurements were taken at this position. This process was repeated until the measured signal could no longer be distinguished from the background noise of the sensor due to the distance. The result is that the measuring system has an axial measuring range of 2.3 mm and is therefore large enough to measure the defined structures in the process. In a next step of the calibration the axial resolution of the measuring system was determined and a correlation between a measured height and the position of a peak was derived. According to DIN 32 877 a random position was approached and the position of the resulting peak of the surface was recorded. A total 50 measuring points within the measuring range were considered. By changing the height of the piezo stepper, the position of the peak in the evaluation domain is changed so that a calibration factor can be determined. By dividing the respective difference between the current and previously approached measuring point Z and the corresponding peak positions P, a factor can be calculated. The arithmetic mean is then calculated over the total number of measurements N, see Figure 5.

$$\frac{1}{N}\sum_{n=1}^{N}\frac{|Z_n - Z_{n-1}|}{|P_n - P_{n-1}|},\tag{4}$$

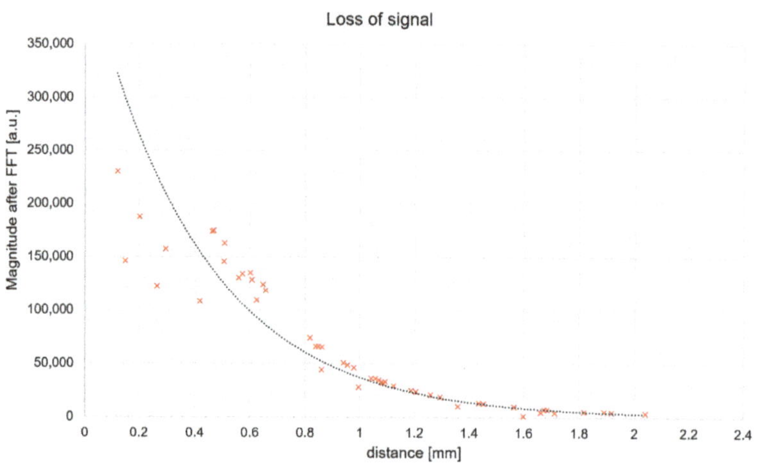

Figure 5. Loss of the measuring signal with increasing measuring distance. The magnitude of the signal decreases with increasing measuring distance until it can no longer be distinguished from the background noise. The red crosses represent one of the random meas.

The result of this calculation is the experimental determination of the axial resolution of the measuring system for a boundary layer. On the basis of the measurement performed, the following results are obtained for the system: A smallest possible distinguishable step size of 3.632 µm with a local repeatability of 0.36 µm. The calculated axial resolution of the system is 3.318 µm with the parameters of the light source used. The real system can only solve structures that are about 10% larger than those determined in advance. Again, it should be mentioned that this case is valid for tomographic structures, which are not considered here. Despite this deviation, the system has a local repeatability of 0.36 µm and is able to resolve the topographical structures required in the project, like mentioned above. Determining a relationship between the tomographic resolution and the repeatability is part of current research activities.

With the offline laboratory system, it could be shown that the developed measuring system is able to resolve the necessary structures and that the interaction of all components works. In the next step, the upscaling for the actual target application, the laser processing system, can take place.

4. Upscaling to Real World Components

As described above, a different wavelength is used for the measurement system than the design wavelength of the optics. This results in a spot offset in the focal plane, since the objective lens used here is not designed for a wavelength nearly 100 nm different from the process laser. As shown in Figure 6, the positioning error increases the further the beam is deflected. For the lens considered here, the maximum spot offset is 171.8 µm. No direct measurement of the laser ablation is possible. Now it is possible to adjust the machining process to correct the offset. This requires an interruption of the machining process and a repositioning of the scanner mirrors, which would lead to a significant slowdown of the entire process. Another possibility would be to determine a correction file for the scanner objective lens system and the measuring wavelength, as it is used for image field calibration with the process laser. However, this would require significant changes to the process. In addition to the actual processing and the necessary extension for synchronization with the measuring system, a change of the correction data and a repositioning of the scanner would have to take place for each individual measurement in order to be able to measure the same location.

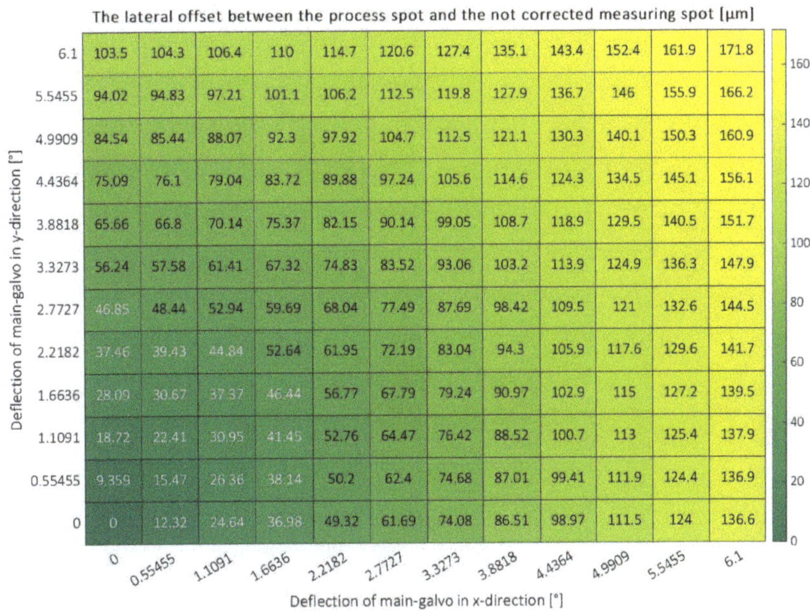

Deflection y-dir [°]	0	0.55455	1.1091	1.6636	2.2182	2.7727	3.3273	3.8818	4.4364	4.9909	5.5455	6.1
6.1	103.5	104.3	106.4	110	114.7	120.6	127.4	135.1	143.4	152.4	161.9	171.8
5.5455	94.02	94.83	97.21	101.1	106.2	112.5	119.8	127.9	136.7	146	155.9	166.2
4.9909	84.54	85.44	88.07	92.3	97.92	104.7	112.5	121.1	130.3	140.1	150.3	160.9
4.4364	75.09	76.1	79.04	83.72	89.88	97.24	105.6	114.6	124.3	134.5	145.1	156.1
3.8818	65.66	66.8	70.14	75.37	82.15	90.14	99.05	108.7	118.9	129.5	140.5	151.7
3.3273	56.24	57.58	61.41	67.32	74.83	83.52	93.06	103.2	113.9	124.9	136.3	147.9
2.7727	46.85	48.44	52.94	59.69	68.04	77.49	87.69	98.42	109.5	121	132.6	144.5
2.2182	37.46	39.43	44.84	52.64	61.95	72.19	83.04	94.3	105.9	117.6	129.6	141.7
1.6636	28.09	30.67	37.37	46.44	56.77	67.79	79.24	90.97	102.9	115	127.2	139.5
1.1091	18.72	22.41	30.95	41.45	52.76	64.47	76.42	88.52	100.7	113	125.4	137.9
0.55455	9.359	15.47	26.36	38.14	50.2	62.4	74.68	87.01	99.41	111.9	124.4	136.9
0	0	12.32	24.64	36.98	49.32	61.69	74.08	86.51	98.97	111.5	124	136.6

Figure 6. Lateral offset between the process spot and the not corrected measuring spot in dependency of the deflection of the two main-galvanometers.

As shown in Figure 7, the spot offset to the border of the objective lens increases. This non-linear or non-constant offset cannot be compensated by additional optical elements, for example, so that a dynamic control is necessary. This dynamic control was made possible by adding another scanner exclusively for the measurement system. As illustrated in Figure 7. At the same time, this opens up new possibilities for process monitoring. It is not only possible to inspect the ablation point, but also a upstream or downstream inspection of the workpiece while the process is running.

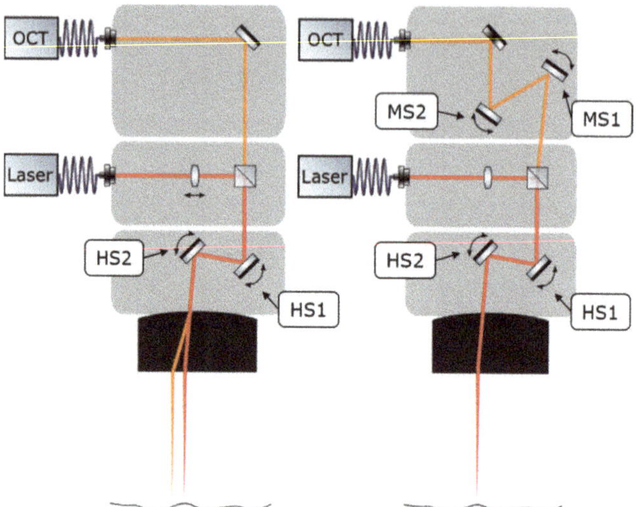

Figure 7. System comparison with and without beam manipulation. Left: the coupling of both beam paths without the adjustment of the measuring beam and the resulting lateral offset on the processing plane; right: an additional pair of galvanometer mirrors (MS1 and MS2).

5. Spot Shift Compensation Model

The possibility to manipulate the spot position of the measuring system alone is not sufficient to achieve synchronous operation of both systems involved in a production environment. To dynamically adjust the spot position of the measuring system so that overlaying between the process laser and the measuring beam occurs, it is necessary to know the tilting angle of the scanner mirrors required for correction.

The optical system was recreated as a Zemax OpticStudio simulation and simulated for the outermost point in the image area of the lens, like shown in Figure 8. If both mirrors of the processing scanner (HS1, HS2 in Figure 7) are deflected to the maximum possible angle of [6.0°,6.0°] for the lens-scanner combination, the result is a sport offset of 171.8 µm. Now the mirrors (MS1 and MS2) are varied in 0.001° steps until the optimum is found. As a result, the spot offset in the simulation could be corrected to 1.8 µm or by 98.9%. Thus, a correction on this basis is possible.

Figure 8. Beam path of the measuring beam with the two main-galvanometer mirrors and the two secondary-galvanometer mirrors.

However, manual correction is very labor- and time-intensive.

It is difficult to determine all possible combinations of angles and resolutions in advance. A simulation algorithm was developed to obtain correction data for arbitrary discretization, hereby, the density of the generated data point on the surface or the amount of data can be influenced. From the simulation of the light beam path, it is known that the measuring spot is smaller in the focus plane than the processing spot. This is sufficient as a boundary condition for the termination condition of the simulation that the spots cover each other and not that the centers must match.

The aim of the data obtained from the simulation model is to generate a correction table or lookup table in which the angles to be set are dependent on the position of the main scanner. At the same time, the amount of data should be as small as possible. On the one hand, this saves time when generating the data, and on the other hand, it reduces the amount of memory required. For this purpose, the scan field is first assumed to be symmetrical, i.e., the values of one quadrant are transferable to the others, thus reducing the data requirement by 75%. The remaining scan field is now discretized in x- and y-direction, whereby the simulation model enables a flexible and independent discretization of the two directions. The actual calculation of the correction values is now carried out on the points created in this way. In the process, the above-mentioned Zemax OpticStudio simulation was extended so that it can run automatically. Several cycles of the simulation are run through. With each cycle, the step size of the angular iteration is reduced until it reaches the mechanical resolution of the galvanometer used. This procedure is illustrated in Figure 9.

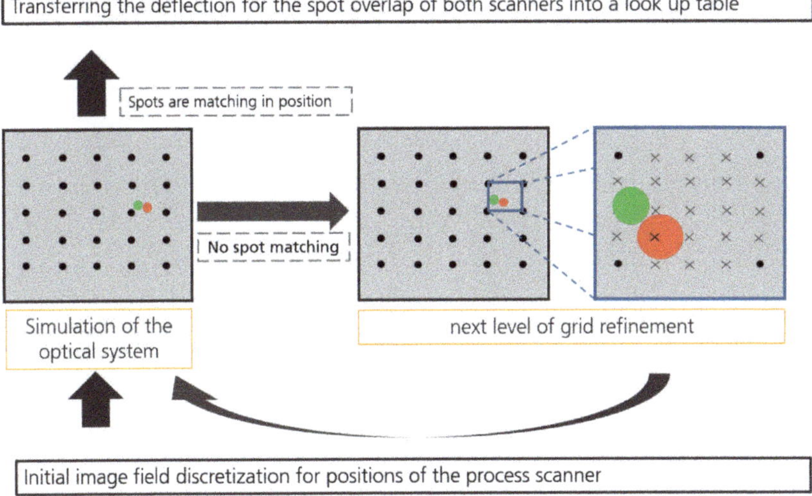

Figure 9. Illustration of the iterative procedure for determining the correction values of the lookup table. Simulations are performed in various degrees of refinement until a spot coverage is achieved.

With the data generated in this way, compensation can take place, but due to the discretization, the grid has gaps. These gaps can be closed by interpolating the data. The laser processing system informs the measuring system of the current position of the main scanner and thus the position at which the measurement is to be made. Scanning systems do not usually work directly with the applied angles, unlike our simulation data. Therefore, the first step here is homogenization. Subsequently, the next interpolation points that are located at the adjacent position are identified and the required values are interpolated. The mechanical resolution of the galvanometer used is also taken into account and, if the tolerance is exceeded, the calculation is carried out again with modified interpolation points. As a result, the angles to be created for the scanning system of the measuring system are now available. Now the actual measurement can take place.

During the calibration of the measuring system, the data for the correction table is calculated and transformed into a coordinate system suitable for the scanner used. Thus, the calculation effort during the actual process is limited to the interpolation of interpolation points based on the correction table. Communication between the machining process and the measuring system is required so that the results can be taken, and the spot position of the measuring system can be corrected. LCI as a point-based method requires triggering of each individual acquisition. For this purpose, the machining process must be adapted so that a measurement is triggered at the desired acquisition time, e.g., by a TTL signal, and at the same time it is necessary that there is no movement of the scanner mirrors at the time of the measurement. In order to correct the sport offset, the current position of the processing scanner must be communicated to the measurement system as an extension of the classical LCI. Then the corresponding value for compensation can be applied from the correction model to the additional galvo pair and the actual measurement can take place. These two steps must be integrated into the actual machining process for the locations for which a measurement is to be made so that accurate results can be obtained.

The simulation model is designed in such a way that optical or mechanical components can be exchanged, e.g., for adaptation to a different setup.

New challenges arise here during integration into the laser machine. In order to be able to accommodate the second scanning system in the laser machine, it is necessary to develop an integration box, shown in Figure 10. This is used to couple the processing laser into the beam splitter and at the same time the scanner unit for the measuring system is housed here and also coupled into the beam path of the machine via the beam splitter. Only the beam splitter has been integrated as a new component in the beam path of the process laser and it has only the influence of a mirror.

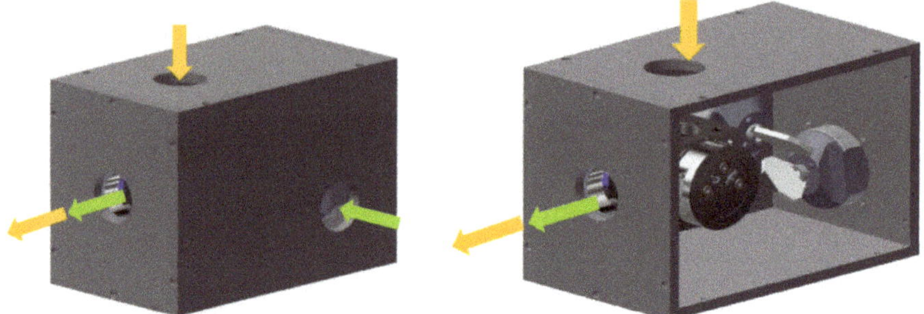

Figure 10. Integration unit for coupling the beam paths in the machine and for manipulating the measuring beam.

This is used to couple the processing laser into the beam splitter and at the same time the scanner unit for the measuring system is housed here and also coupled into the beam path of the machine via the beam splitter. Only the beam splitter has been integrated as a new component in the beam path of the process laser.

6. Discussion and Conclusions

The measurement system based on low coherence interferometry described here enables measurements to be taken directly at the point of ablation. This also requires a new method of controlling the machining process. The necessary synchronization between the machining process and the measurement increases the machining time for a workpiece. The increase in time depends on factors such as the number of measuring points or the possible recording frequency of the spectrometer used and is therefore dependent on the actual process. No removal of the workpiece is necessary, so there is no need for repositioning work. The ablation of a laser structuring process usually takes place in several cycles, with material in the nm range being ablated in each cycle [32]. This is also a limitation of the LCI process, with a repeatability of 0.36 µm an ablation cycle is not detectable, but a set of them. If the workpiece is inspected at the end of the machining process to determine target parameters or to adjust process parameters, the resulting delay is much smaller than if each cycle of the process is measured. If the machining process runs without the use of the measuring system, there is no delay or further influence on the process.

The basic challenge of the spot offset was solved by integrating a separate scanner for the measuring system and a model for control. This integration also opens up new application possibilities. In this context, the goal was to achieve a spot overlap, which was successfully achieved with the control model. The model can be adapted to other optical systems. The generation of the correction data is very complex. There is a need for further research to reduce the amount of data and to find a more effective way to integrate the control into the process [33]. The independent control of the measuring and processing beam allows further investigations of the measuring strategy. For example, the measuring spot can be positioned in front of or behind the ablation point and new process data can be generated or interactions between the measuring light and the plasma created by the ablation can be bypassed. In addition to pure process monitoring, data volumes for AI approaches can also be recorded in this way. Spot offset compensation can be useful not only in the topographical case discussed here, but also in the application of classical OCT, e.g., for process monitoring in laser transmission welding [34]. However, further research is needed on these points.

With the methods presented here, an extension of the LCI by a sport offset compensation was made and the system was developed to meet the specific project requirements. It could be shown that the target structures can be solved, and how a spot offset can be compensated, or which mechanical adjustments have to be made for a system integration into an existing machine. As the ESSIAL research project is still ongoing, the presented method will be further developed and adapted if necessary.

Author Contributions: Conceptualization, F.Z.; methodology, F.Z. and J.J.; software, F.Z. and J.J.; validation, F.Z. and J.J.; formal analysis, F.Z.; investigation, F.Z.; resources, F.Z.; data curation, F.Z. and J.J.; writing—original draft preparation, F.Z. and J.J.; writing—review and editing, F.Z., J.J. and R.H.S.; visualization, F.Z. and J.J.; supervision, R.H.S.; project administration, F.Z. and R.H.S.; funding acquisition, R.H.S. All authors have read and agreed to the published version of the manuscript.

Funding: The research was funded by European Commission Project ESSIAL (Grant Agreement n. 766437, EU H2020 RIA FOF-06-2017).

Conflicts of Interest: The authors declare no conflict of interest.

Appendix A

Table A1. Nomenclature.

Symbol	Definition	Units
σ_z	axial resolution	µm
$\sigma_{(x,y)}$	lateral resolution	µm
$\Delta\lambda$	spectral width (FWHM)	nm
λ_0	central wavelength	nm
N	number of pixels	-
n	refractive index	-
f	focal length	mm
d	diameter of the collimated beam diameter at objective lens entry	mm
(P_n, Z_n)	pair of values of peak position and distance	(-,mm)

References

1. Yabumoto, M.; Kaido, C.; Wakisaka, T.; Kubota, T.; Suzuki, N. *Electrical Steel Sheet for Traction Motorsof Hybrid/Electric Vehicles*; Nippon Steel: Tokyo, Japan, 2003.
2. Schade, T.; Ramsayer, R.M.; Bergmann, J.P. Laser welding of electrical steel stacks investigation of the weldability. In Proceedings of the 4th International Electric Drives Production Conference (EDPC), Nuremberg, Germany, 30 September–1 October 2014.
3. Sievert, J. The measurement of magnetic properties of electrical sheet steel—Survey on methods and situation of standards. *J. Magn. Magn. Mater.* **2000**, *215*, 647–651. [CrossRef]
4. Heilemann, S.; Zwahr, C.; Knape, A.; Zschetzsche, J.; Lasagni, A.F.; Füssel, U. Improvement of the Electrical Conductivity between Electrode and Sheet in Spot Welding Process by Direct Laser Interference Patterning. *Adv. Eng. Mater.* **2018**. [CrossRef]
5. Lahn, L.; Wang, C.; Allwardt, A.; Belgrand, T.; Blaszkowski, J. Improved Transformer Noise Behavior by Optimized Laser Domain Refinement at ThyssenKrupp Electrical Steel. *IEEE Trans. Magn.* **2012**, *48*, 1453–1456. [CrossRef]
6. Patri, S.; Gurusamy, R.; Molian, P.A.; Govindaraju, M. Magnetic domain refinement of silicon-steel laminations by laser scribing. *J. Mater. Sci.* **1996**, *31*, 1693–1702. [CrossRef]
7. Pérez-Belis, V.; Bovea, M.D.; Ibáñez-Forés, V. An In-Depth Literature Review of the Waste Electrical and Electronic Equipment Context: Trends and Evolution. *Waste Manag. Res.* **2014**, *33*, 3–29. [CrossRef] [PubMed]
8. Bärsch, N.; Körber, K.; Ostendorf, A.; Tönshoff, K.H. Ablation and cutting of planar silicon devices using femtosecond laser pulses. *Appl. Phys. A* **2003**, *77*, 237–242. [CrossRef]
9. Gillner, A.; Gretzki, P.; Büsing, L. High power parallel ultrashort pulse laser processing. In Proceedings of the SPIE LASE, San Francisco, CA, USA, 13–18 February 2016.
10. Maloberti, O.; Meunier, G.; Kedous-Lebouc, A. Hysteresis of Soft Materials Inside Formulations: Delayed Diffusion Equations, Fields Coupling, and Nonlinear Properties. *IEEE* **2008**, *44*, 914–917. [CrossRef]
11. Maloberti, O.; Kedous-Lebouc, A.; Meunier, G.; Mazauric, V. How to formulate soft material heterogeneity? I. Hysteresis dynamic motion and behaviour. In Proceedings of the Soft Magnetic Materials Conference—SMM18, Cardiff, UK, 3 September 2007.
12. Majumdar, J.D.; Manna, I. Laser processing of materials. *Sadhana* **2003**, *28*, 495–562. [CrossRef]
13. Schmitt, R.; Mallmann, G.; Winands, K.; Pothen, M. Inline Process Metrology System for the Control of Laser Surface Structuring Processes. *Phys. Procedia* **2012**, *39*, 814–822. [CrossRef]
14. Neiheisel, G.L. Full Production Laser Processing Of Electrical Steel. In *Full Production Laser Processing of Electrical Steel*; SPIE: Washington, DC, USA, 1986; pp. 116–126.
15. Seo, U.J.; Kim, D.J.; Chun, Y.D.; Han, P.W. Mechanical cutting effect of electrical steel on the performance of induction motors. *Energies* **2020**, *13*, 6314. [CrossRef]
16. Matsumura, K.; Fukuda, B. Recent developments of non-oriented electrical steel sheets. *IEEE Trans. Magn.* **1984**, *20*, 1533–1538. [CrossRef]
17. Kurosaki, Y.; Mogi, H.; Fujii, H.; Kubota, T.; Shiozaki, M. Importance of punching and workability in non-oriented electrical steel sheets. *J. Magn. Magn. Mater.* **2008**, *320*, 2474–2480. [CrossRef]
18. Oda, Y.; Kohno, M.; Honda, A. Recent development of non-oriented electrical steel sheet for automobile electrical devices. *J. Magn. Magn. Mater.* **2008**, *320*, 2430–2435. [CrossRef]
19. Klocke, F.; König, W. *Abtragen, Generieren und Lasermaterialbearbeitung*; Springer: Berlin/Heidelberg, Germany, 2007.
20. Ji, Y.; Van Vlack, C.; Webster, P.J.; Fraser, J.M. Real-Time Depth Monitoring of Galvo-telecentric Laser Machining by Inline Coherent Imaging. In Proceedings of the CLEO OSA Technical Digest, San Jose, CA, USA, 9–14 June 2013.
21. Schmitt, J.M. Optical coherence tomography (OCT): A review. *IEEE J. Sel. Top. Quantum Electron.* **1999**, *5*, 1205–1215. [CrossRef]
22. Brezinski, M. *Optical Coherence Tomography: Principles and Applications*; Elsevier: Amsterdam, The Netherlands, 2006.
23. Song, G.; Harding, K. *OCT for Industrial Applications*; SPIE: Bellingham, WA, USA, 2012.
24. Meyer, H.; Zabic, M.; Kaierle, S.; Ripken, T. Optical coherence tomography for laser transmission joining processes in polymers and semiconductors. *Procedia CIRP* **2018**, *74*, 618–622. [CrossRef]
25. Tomlins, P.H.; Wang, R.K. Theory, developments and applications of optical coherence tomography. *J. Phys. D Appl. Phys.* **2005**, *38*, 2519. [CrossRef]
26. Schmitt, R.; Mallmann, G.; Peterka, P. Development of a FD-OCT for the inline process metrology in laser structuring systems. In Proceedings of the SPIE Optical Metrology, Munich, Germany, 7 December 2011.
27. Kunze, R.; König, N.; Schmitt, R. Monitoring of laser material processing using machine integrated low-coherence interferometry. In Proceedings of the Fifth International Conference on Optical and Photonics Engineering, Singapore, 6 November 2017.
28. Zechel, F.; Kunze, R.; König, N.; Schmitt, R.H. Optical coherence tomography for non-destructive testing. *Tm-Technisch. Mess.* **2020**, *87*, 404–413. [CrossRef]
29. Kunze, R.; Schmitt, R. Inline surface topography measurements of ultrashort laser pulsed manufactured micro structures based on low coherence interferometry. *TM-Technisch. Mess.* **2017**, *84*, 575–586. [CrossRef]
30. Zechel, F.; Kunze, R.; Widmann, P.; Schmitt, R.H. Adaptive process monitoring for laser micro structuring of electrical steel using Optical Coherence Tomography with non-colour corrected lenses. *Procedia CIRP* **2020**, *94*, 748–752. [CrossRef]
31. Kunze, R.; Bredol, P.; Schmitt, R. Signal processing for in-line process monitoring for coaxially integrated metrology in laser micromachining applications. *De Gruyter Oldenbourg* **2017**, *65*, 416–425.

32. Žemaitis, A.; Gaidys, M.; Brikas, M.; Gečys, P.; Račiukaitis, G.; Gedvilas, M. Advanced laser scanning for highly-efficient ablation and ultrafast surface structuring: Experiment and model. *Sci. Rep.* **2018**, *8*, 2045–2322. [CrossRef] [PubMed]
33. Xie, J.; Huang, S.; Duan, Z.; Shi, Y.; Wen, S. Correction of the image distortion for laser galvanometric scanning system. *Opt. Laser Technol.* **2005**, *37*, 305–311. [CrossRef]
34. Schmitt, R.; Ackermann, P. OCT for Process Monitoring of Laser Transmission Welding. *Laser Tech. J.* **2016**, *13*, 15–18. [CrossRef]

Article

Camera-Based in-Process Quality Measurement of Hairpin Welding

Julia Hartung [1,2,*], **Andreas Jahn** [1], **Oliver Bocksrocker** [3] and **Michael Heizmann** [2]

[1] TRUMPF Laser GmbH, Aichhalder Str. 39, 78713 Schramberg, Germany; andreas.jahn@trumpf.com
[2] Institute of Industrial Information Technology, Karlsruhe Institute of Technology, Hertzstraße 16, 76187 Karlsruhe, Germany; michael.heizmann@kit.edu
[3] TRUMPF GmbH + Co. KG, Johann-Maus-Str. 2, 71254 Ditzingen, Germany; oliver.bocksrocker@trumpf.com
* Correspondence: julia.hartung@trumpf.com

Abstract: The technology of hairpin welding, which is frequently used in the automotive industry, entails high-quality requirements in the welding process. It can be difficult to trace the defect back to the affected weld if a non-functioning stator is detected during the final inspection. Often, a visual assessment of a cooled weld seam does not provide any information about its strength. However, based on the behavior during welding, especially about spattering, conclusions can be made about the quality of the weld. In addition, spatter on the component can have serious consequences. In this paper, we present in-process monitoring of laser-based hairpin welding. Using an in-process image analyzed by a neural network, we present a spatter detection method that allows conclusions to be drawn about the quality of the weld. In this way, faults caused by spattering can be detected at an early stage and the affected components sorted out. The implementation is based on a small data set and under consideration of a fast process time on hardware with limited computing power. With a network architecture that uses dilated convolutions, we obtain a large receptive field and can therefore consider feature interrelation in the image. As a result, we obtain a pixel-wise classifier, which allows us to infer the spatter areas directly on the production lines.

Keywords: hairpin; laser welding; semantic segmentation; dilated convolution; sdu-net; spatter detection; quality assurance; fast prediction time

Citation: Hartung, J.; Jahn, A.; Bocksrocker, O.; Heizmann, M. Camera-Based in-Process Quality Measurement of Hairpin Welding. *Appl. Sci.* **2021**, *11*, 10375. https://doi.org/10.3390/app112110375

Academic Editor: Andreas Fischer

Received: 13 September 2021
Accepted: 19 October 2021
Published: 4 November 2021

Publisher's Note: MDPI stays neutral with regard to jurisdictional claims in published maps and institutional affiliations.

Copyright: © 2021 by the authors. Licensee MDPI, Basel, Switzerland. This article is an open access article distributed under the terms and conditions of the Creative Commons Attribution (CC BY) license (https://creativecommons.org/licenses/by/4.0/).

1. Introduction

In the production of electric motors, the automotive industry relies on a new technique known as hairpin. Instead of twisted copper coils, single copper pins that are bent like hairpins are used, which give the technology its name. These copper pins are inserted into the sheet metal stacks of a stator and afterward welded together in pairs. As with conventional winding, the result is a coil that generates the necessary magnetic field for the electric motor. This method replaces complex bending operations and enables a more compact motor design while saving copper material [1,2]. Depending on the motor design, between 160 and 220 pairs of hairpins per stator are welded. If at least one weld is defective, the entire component may be rejected. Therefore, continuous quality control is necessary and the weld of every hairpin pair should be monitored [1].

In most cases, a laser is used to weld the pins. The laser welding process enables a very specific and focused energy input, which ensures that the insulation layer is not damaged during the process. In addition, unlike electron beam welding, no vacuum is required and laser welding is a flexible process that can easily be automated in a short cycle time [1]. The lower-cost laser sources that are scalable in the power range emit in the infrared wavelength range, which is comparatively difficult for working with copper. At this wavelength of about 1030 nm or 1070 nm copper, is highly reflective at room temperature, so very little incoming laser light is absorbed [1,3]. Just before reaching the melting temperature the absorption level rises from 5% to 15% and reaches almost 100%

when the so-called keyhole is formed. Based on this dynamic, the process is prone to defects and spattering [4]. A spatter occurs when the keyhole closes briefly and the steam pressure causes the material to leak out of the keyhole. If the ejected material gets into the stator, it may cause short circuits or other defects [1]. In addition, less material will be used to form the weld, which often leads to a loss of stability. For these reasons, it is extremely important to prevent spatter as much as possible. Various processes can improve the welding result on copper. Three approaches are briefly touched upon below. By moving the laser spot fast and simultaneously during forwarding motion (wobbling), stable dynamics can be created in the weld pool. This can improve the process quality when welding with an infrared laser. Another approach is welding with different strengths of inner and outer fiber core. This means that the inner fiber core is used to create the desired welding depth with high intensity, while the molten pool is stabilized by an outer fiber core—the fiber ring. In addition, there is the possibility of using a visible wavelength of a green laser, which results in higher absorption of the laser light and thus higher process reliability [5–7]. Furthermore, there may also be external causes that lead to spattering. These include, for example, contamination, gaps, misalignment or an oxidized surface.

The correct setting of the laser welding parameters such as laser power, speed and focus size is very important in copper welding. In addition, the process may not drift or this must be detected at an early stage. The presence of spatter on the component can be used as an indicator of an unstable situation in the welding process, as its occurrence is closely related to the quality of the weld seam [8,9]. Due to the briefly mentioned reasons, it is essential to monitor the welding process while focusing particularly on spattering. This allows a conclusion about the quality of individual welds, the occurrence of defects, as well as the overall quality of the stator. An important requirement is also fast process time which is a prerequisite for a system to be used in large-scale production. The welding of an entire engine takes just a bit more than one minute and quality monitoring should not slow down the process [1,2].

Currently, there are only a few machine learning applications that are used for quality assessment in laser welding [10]. Some approaches are presented by Mayr et al. [11], including an approach for posterior quality assessment based on images using a convolutional neural network (CNN). They use three images in the front, back, and top view of a hairpin, to detect quality deviations [12]. In [13] a weld seam defect classification with help of a CNN is shown. They achieve a validation accuracy of 95.85% in classifying images of four different weld defects, demonstrating the suitability of CNNs for defect detection. Nevertheless, some defects cannot be seen visually on the cooled weld seam. For example, pores in the weld seam or a weld bead that is too small due to material ejection can not be visually distinguished from a good weld seam.

That is why imaging during the process of hairpin welding offers more far-reaching potential for machine learning than subsequent image-based inspection of the weld. Important criteria are the mechanical strength of the pin and the material loss [12]. Both criteria are in correlation with a stable welding process and the occurrence of spatter. For spatter detection, a downstream visual inspection of the component is also possible [14]. However, this approach is problematic for hairpins since there is little material around the hairpins that can be verified for spattering.

Therefore, this paper presents an approach that enables spatter detection during hairpin welding. One of the main challenges of spatter detection directly during the welding process is the fast execution time on hardware with low computing power. The algorithm should be executed directly in the production line, where the installed hardware is often fanless and only passively cooled due to ingress protection. Another important issue is the amount of training data. Since this is an application in an industrial environment, training should only be done on a small data set so that the labeling effort is low and the algorithm can be quickly adapted to new processes. These two aspects are considered in the following.

In Section 2 the data basis and the analysis methods are presented in detail. On the one hand, the network architecture is discussed, but also comparative algorithms, such as morphological filters with their configurations, are presented. Subsequently, in the result section, the training parameters and the results are shown. Finally, the results are discussed and summarized in Section 4.

2. Materials and Methods

To obtain a comprehensive data basis, images were recorded from two different perspectives while welding hairpins. The captured images were then used to perform different approaches to monitor the occurrence of spatter. An automated solution for spatter detection is recommended because it ensures consistent evaluation and only in this automated way can spatter be detected reliably. The image-based approach contributes to the detection of spatter caused by external factors. By using artificial intelligence, a high feature variance in the data is covered. Additionally, the required properties of the categories do not need to be defined in detail. Images were continuously recorded in the process to implement in-process quality monitoring. Since a visual inspection of the hairpins after the welding process is not always clear, a post-weld inspection is not precise enough.

2.1. Data Basis

We use three different approaches for generating data sets for the observation of the welding process using an industrial camera. These differ firstly in the perspective from which the data were recorded and secondly in the type of preprocessing.

In the first approach, the images taken during the welding process were summed up to a single image per process. The superimposition of the individual images provides information about the spatter occurrence during the entire process [15] and can thus be used to evaluate the welding behavior. When summing up the images, those taken at the time the laser beam pierced the material must be removed. In this process step, a white glow occurs, which takes up so much space on the image, that it would hide possible spatters. An example is shown in Figure 1.

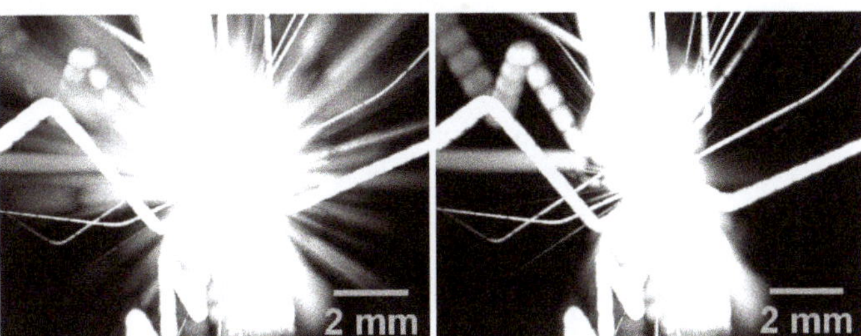

Figure 1. Summed image with and without piercing.

On the one hand, the images were taken with a high-speed camera mounted on the side of the welding process. This provides a lateral view in which the welding process and spatter are visible. The data set containing the summed images of a welding process is referred to as *laterally complete* in the following.

The second view of the images is coaxial through the laser optics. Often, a camera is already installed in a laser welding system that captures images through laser optics. These images are used for example to determine the component position before the welding process. To achieve more accurate results in this process, the magnification optics are often selected to provide a good imaging ratio for the component. In contrast, when monitoring the welding process for spatter, the lowest possible magnification would be optimal. Since the camera is usually part of an existing system, the magnification cannot be adjusted specifically, which often means that a higher magnification must be used. For our test setup, we use a gray-scale camera with a recording frequency of 2 kHz. The summed images, which were generated based on the coaxial view, are called *coaxial complete*.

For a third approach, we evaluate the individual images acquired by the coaxial view. The third data set is called *coaxial single*. While spatters are shown as lines on the summed image, they are usually visible as dots on the single images, depending on the exposure time.

With semantic segmentation, labeling is time-consuming and error-prone. Many developments in the industry are carried out specifically for a customer project on customer data. The labeling effort, which means time resources and therefore costs, is recurring for each customer project. In addition, there are often confidentiality agreements and data flow for a large database is difficult. Therefore, especially in the industry, an attempt should be made to work with a small amount of training data. To teach the network the desired invariance and robustness properties even in training with only a few training data sets, data augmentation is essential [16]. Various network architectures, such as the U-Net architecture, are designed for strong data augmentation. We enlarge our training data set using rotation, vertical and horizontal shift, vertical and horizontal flip, adjustment of the brightness range, zoom and shear. Through the strong use of data augmentation we also have the advantage of avoiding overfitting. Dosovitskiy et al. [17] have shown in this context the importance of data augmentation to learn a certain invariance. Despite a small data set, the network never gets the same image with an identical setting presented multiple times. Thus, it cannot learn the image by memory.

The segmentation masks could not be exact for each pixel. Especially in the summed images where the spatter is shown as a line, the labels are not accurate for each pixel. As shown by Tabernik et al. [18] inaccurate labeling is sufficient for predictions aimed at a quality assurance based on defect detection. The paper processes the topic of surface-defect detection. They used a segmentation-based approach to detect cracks on an image and transferred the problem afterwards to binary image classification with the classes *defect is present* and *defect is not present*. In an experiment, it was shown that the segmentation results are better if larger areas around the crack are marked in the annotation mask. Although pixels that are not part of the crack are marked as defect class, they achieve better results than with an accurate annotation mask. Our use case is very similar to the use case of Tabernik et al., which is why their result can be applied to our task. Instead of surface defects, we detect spatter in our application.

We train separate models, one for each data generation approach. For the two models based on the summed images, we use 14 images each in the training process. Each of these images represents the complete welding process, whereas with the *coaxial single* data set we have about 700 images per welding process. There, we use 500 images from different processes for training. We categorize the pixel-wise class assignment into the background, process lights, and spatter.

2.2. Network Architecture

The images are evaluated using a neural network. We use a segmentation network to localize the process light and the spatter pixel by pixel. Compared to the object detection, which could be done for example with YOLO [19] or SSD [20], this has the advantage that through the pixel-based loss function each pixel can be considered as an individual training instance [18]. This increases the effective number of training data massively and thus also counteracts overfitting during training.

Unlike many other network architectures, the U-Net architecture is very well suited for small data sets, as shown by Ronneberger et al. [16]. Therefore a neural network whose architecture is based on the stacked dilated U-Net (SDU-Net) architecture presented by Wang et al. [21] is used to evaluate our images. The abbreviation SD is short for stacked dilated. To avoid confusion considering the SDU-Net, it should be mentioned that various other U-Net modifications exist, which are also called SDU-Net. However, in these nets the abbreviation SD describes other concepts. For example, there is the spherical deformable

U-Net (SDU-Net) from Zhao et al., which was developed for medical imaging in the inherent spherical space [22]. Because there is no consistent neighborhood definition in the cortical surface data, they developed another type of convolution and pooling operations, especially for this data. Another U-net modification, which is also called SDU-Net, is the structured dropout U-Net presented by Guo et al. [23]. Instead of the traditional dropout for convolutional layers, they propose a structured dropout to regularize the U-Net. Gadosey et al. present the stripping down U-Net, with the same abbreviation, for segmentation images on a platform with low computational budgets. By use of depth-wise separable convolutions, they design a lightweight deep convolutional neural network architecture inspired by the U-Net model [24].

As mentioned before, we used a SDU-Net modification with stack dilated convolutional layers. This U-Net variant adopts the architecture of the vanilla U-Net but uses stacked dilated convolutions. Instead of using two standard convolutions in each encoding and decoding operation, the SDU-Net uses one standard convolution followed by multiple dilated convolutions which are concatenated as input for the next operation. Thus the SDU-Net is deeper than a comparable U-Net architecture and has a larger receptive field [21]. In Figure 2 our architecture is shown in detail. We used a gray-scale image of the size 256×256 pixels as input. In designing the network architecture we set the concatenate output channel numbers to $n_1 = 16$, $n_2 = 32$, $n_3 = 64$ and $n_4 = 128$ instead of $n_1 = 64$, $n_2 = 128$, $n_3 = 256$ and $n_4 = 512$ like the original implementation of the paper. Since our images are far less complex compared to the medical images used in the paper from Wang et al., this number is sufficient. So all in all we have 162,474 trainable parameters instead of 6,028,833. With this comparatively small amount, we achieve a fast inference time, which is about 20 ms on CPU. This is important to be able to run the network prediction on an industrial computer directly at the production line, where often no GPU is available.

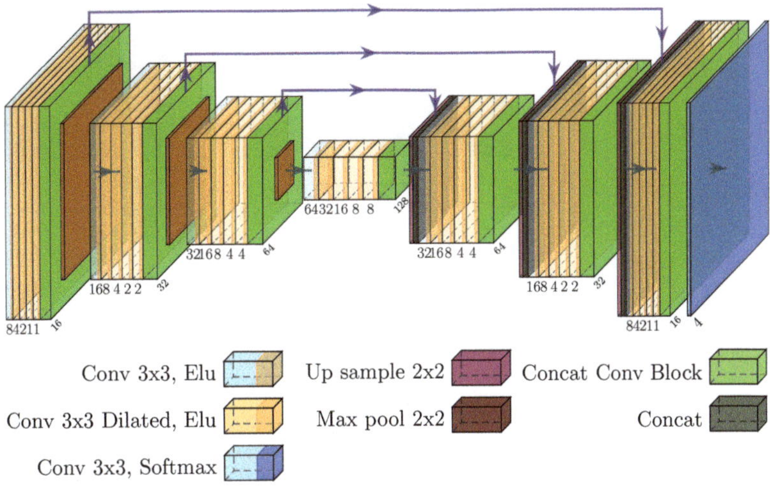

Figure 2. Our optimized small SDU-Net architecture. The input size is 256×256 pixels and we have 32×32 pixels at the lowest resolution. The number of channels is denoted at the bottom of the boxes.

2.3. Evaluation

An alternative approach is to implement spatter detection with a morphological filter. We choose the opening operation, which involves an erosion of the data set I followed by dilation, both with the same structural element H:

$$I \circ H = (I \ominus H) \oplus H. \tag{1}$$

In the first step, the erosion, the opening process eliminates all foreground structures that are smaller than the defined structural element. Subsequently, the remaining structures are smoothed by the dilation and thus grow back to approximately their original size. With the opening filter, we identify the process light in the images based on the structure element H. In Figure 3 the process is visualized. An input image is shown in Figure 3a and the corresponding filtered image in Figure 3b. Afterwards, we define the spatters by subtracting the filtered image from the original image. The remaining image elements represent the spatters, shown in Figure 3c. Figure 3d shows an overlaid image, where the process light is painted in green and the spatters in red. For this algorithm we use the original image size with 480×640 pixels for the coaxial images and 840×640 pixels for the lateral images. We choose the structure elements as ellipse with $H = 45 \times 45$ pixels for the *coaxial single*, $H = 90 \times 90$ pixels for *coaxial complete*, and $H = 60 \times 60$ pixels for *lateral complete* images. The definition of the structural element is based on the average size of the process light, which is estimated in the images. The spatters usually represent smaller elements and can thus be distinguished from the elements found by the filter.

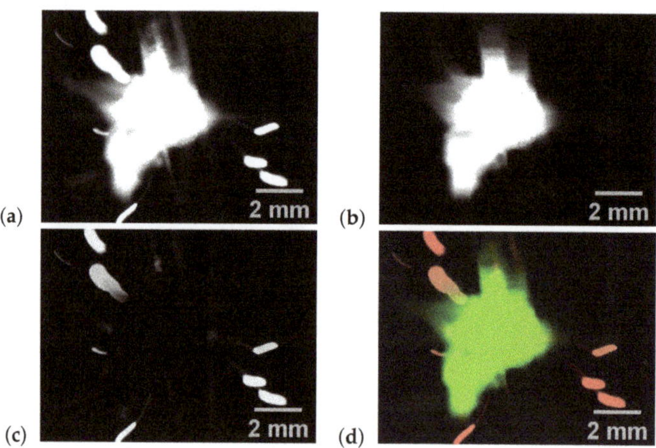

Figure 3. Gray-scale opening, (**a**) gray-scale image (**b**) process light (**c**) spatter (**d**) overlaid image with spatters colored in red and the process light in green.

We have also used other network models, such as a comparable small version of the U-Net model according to Ronneberger et al. [16]. We trained this model equivalent to the SDU-Net architecture with the same input images and the same parameters, shown in the next section.

3. Results

We trained different models of the small SDU-Net architecture for each input data generation approach, *coaxial single*, *coaxial complete*, and *lateral complete*. All models were trained with a batch size of 6, an input of gray-scale images in the size of 256×256 pixels, and 500 steps per epoch. We used the Adam Optimizer and started the training process with a learning rate of 0.001. The learning rate was reduced by 5% after 3 epochs without any improvement until a learning rate of 0.000005 is reached. The training process was stopped when no further improvement has occurred in 20 consecutive epochs. This results in different long training times for the different models. The loss value and the accuracy of the different models can be seen in Table 1. To verify the results during training, we used validation data sets. These contained 3 images each for *coaxial complete* and *lateral complete* and 18 images for *coaxial single*, according to the small database. The validation data sets were also enlarged with strong use of data augmentation. After the training, we used a separate test data set, each containing 50 images with the corresponding ground truth image.

Because the number of pixels per class is very unbalanced, and especially the less important background class contains the most pixels, we used the loss functions weighted dice coefficient loss (DL) and the categorical focal loss (FL) [25]. The network results are shown in Figure 4 and in Table 2.

The advantage of focal loss is that no class weights have to be defined. The loss function, which is a dynamically scaled cross entropy (CE) loss, down-weights the contribution of easy examples and focuses on learning hard examples:

$$FL(p_t) = -\alpha(1 - p_t)^\gamma log(p_t) \tag{2}$$

The two parameters α and γ have to be defined. The parameter α represents the balancing factor, while γ is the focusing parameter. The CE loss is multiplied by the factor $(1 - p_t)^\gamma$. This means that with the value $\gamma = 2$ and a prediction probability of 0.9, the multiplier would be 0.1^2, i.e., 0.01, making the FL in this case 100 times lower than the comparable CE loss. With a prediction probability of 0.968, the multiplier would be 0.001, making the FL already 1000 times lower. This gives less weight to the easier examples and creates a focus on the misclassified data sets. With $\gamma = 0$ the FL works analogously to the cross entropy. Here the values $\alpha = 0.25$ and $\gamma = 2$ were chosen.

For comparison, we used the weighted dice coefficient loss, where the loss value is calculated for each class, weighted with the respective class weighting, and then added up. The class weights were calculated based on the pixel ratio of the respective class in the training images. The classes that contain only a few pixel values, such as the spatter class, must be weighted more heavily so they are considered appropriately during training. Since the values are calculated based on the number of pixels in the training data, these weights vary between the different input data sets.

Besides training individual models for each input data approach, we also trained one model for the prediction of all data, the coaxial and lateral view, summed, and single images. Since the different data sets have the same classes and a similar appearance, one model approach is also possible. The advantage of this approach is that a higher variance of data can be covered in one model and therefore we do not need to define new models or parameters for each data type. To train the global model we used 14 images of each *coaxial* and *lateral complete* data set and 34 *coaxial single* images.

Another advantage is that additional classes can be added to the model. We have introduced a new class, which includes the cooling process of the welding bead. From the moment the process light turns off, the weld is assigned to the cool-down class. This class cannot be identified via the previously described structure recognition with the subsequent exclusion procedure using the morphological filter. Only image elements of different sizes can be detected and distinguished from each other. For elements of similar size with different properties, the method reaches its limits.

The result of training the small SDU-Net as a single model for all data is also shown in Table 1. All four classes were considered in the training process. In the summed images, the cooling process is not visible. For comparison, a SDU-Net model with twice the number of filters was trained. This net has more trainable parameters, but in our test, no significantly better results in loss, accuracy as well as in evaluation could be obtained. In addition, the results of the comparatively small U-Net model are shown.

The classification results are compared using the Intersection over Union metric (*IoU*). The metric compares the similarity between the predicted classification mask and a drawn annotation mask as ground truth by dividing the size of the intersection by the size of the union:

$$IoU(A, B) = \frac{A \cap B}{A \cup B}. \tag{3}$$

Table 1. Training results of the neural network models based on the SDU-Net architecture. Three different models are trained for each input data generation approach. The fourth model was trained for the prediction of all data. In comparison the results of an SDU-Net architecture with twice the number of filters and of the small U-Net are shown.

	Coaxial Single	Coaxial Complete	Lateral Complete	All Data	SDU-Net Double Filter Size, All Data	U-Net, All Data
Steps per epoch				500		
Batch size				6		
Input size [pixels]				256×256		
DL	0.173	0.196	0.198	0.156	0.146	0.163
Accuracy	0.993	0.966	0.995	0.980	0.980	0.975
Validation DL	0.488	0.347	0.489	0.361	0.339	0.428
Validation Accuracy	0.976	0.831	0.953	0.963	0.963	0.951

In Table 2 the evaluation results of the different approaches are shown. The second value in the rows shows the IoU for all pixels in the entire image. The dark area around the weld is most correctly classified as background. Especially for the coaxial single dataset, the background takes up the main part of the image, so the IoU over the entire image is very high for all methods. Therefore, the specific class pixels are again considered separately. This consideration can be seen in the first values in the table. This represents the IoU based on the pixels assigned to a specific class, except for the background class. This value gives more information about the actual result than the total IoU. However, the larger the area of the background, the fewer pixels are included in the calculation of the class-specific IoU. As result, the value is more influenced by individual misclassified pixels.

In Figures 4 and 5 the first value, without consideration of the background pixels, is used. As shown in Figure 4 the two weighted loss functions, FL and DL, result in comparable distributions in which the DL performs only marginally better.

Using a single model trained on all three data sets, an outlier with IoU close to 0 can be seen in each of our test sets in Figure 4. There, a shot during the cooling process with spatter was misclassified as process light. When using different models per data set, this error case did not occur. On the one hand, the error can be attributed to an underrepresentation of the cooling class in the overall data set, since this only occurs in the *coaxial single* images. On the other hand, the occurrence of spatter on images at this point in the welding process is very rare, which is why the case was not sufficiently present in the training. In productive use, it is assumed that the data is taken only from one perspective. Nevertheless, this experiment can show that the model generalizes well even on different input data with only very few training data and thus covers a high data variance.

Figure 5 shows the IoU without considering the background pixels of the different data sets all trained with the dice coefficient loss. This graph shows that the largest deviations are contained in the *coaxial single* data set. In these images, the background occupies the largest image area, which makes small deviations of the other classes more significant, as shown in Table 3.

In all three input data sets, *coaxial single*, *coaxial complete*, and *lateral complete*, the SDU-Net provides the best results compared to the other methods. The disadvantages of the U-net architecture arise from the fact that only simple and small receptive fields are used, which leads to a loss of information about the image context. In our use case, it leads to the fact that the classes cannot always be clearly distinguished from each other. The SDU-Net processes feature maps of each resolution using multiple dilated convolutions successively and concatenates all the convolution outputs as input to the next resolution. This increases the receptive field and both, smaller and larger receptive fields are considered in the result.

Table 2. Evaluation results of the different approaches. The first value shows the average IoU value of the pixel which was assigned to a specific class, excluding the background class and the second value shows the average IoU value for the entire image.

	Morph. Binary	Morph. Gray	SDU-Net, Different Models, DL	SDU-Net, Different Models, FL	SDU-Net, One Model, DL	SDU-Net, One Model, FL	U-Net, One Model, DL
Coaxial Single	0.406, 0.962	0.421, 0.958	0.819, 0.980	0.821, 0.990	0.746, 0.990	0.739, 0.989	0.709, 0.965
Coaxial Complete	0.731, 0.847	0.722, 0.568	0.830, 0.909	0.810, 0.937	0.850, 0.953	0.840, 0.952	0.828, 0.848
Lateral Complete	0.497, 0.860	0.482, 0.920	0.644, 0.935	0.611, 0.968	0.681, 0.967	0.688, 0.984	0.659, 0.932
Average	0.544, 0.889	0.542, 0.789	0.764, 0.943	0.747, 0.965	0.759, 0.972	0.756, 0.975	0.732, 0.915

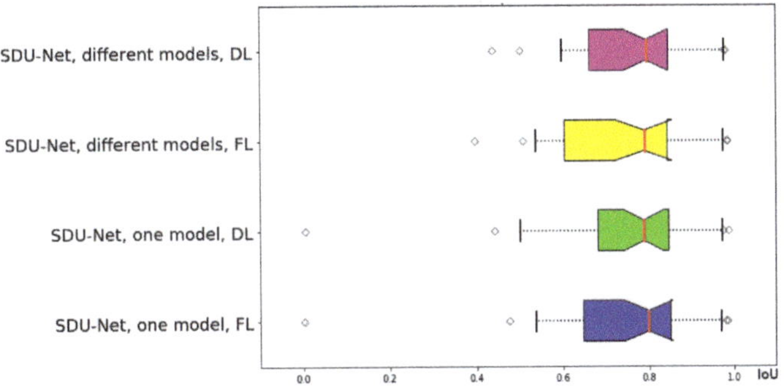

Figure 4. Comparison of the IoU of the classes process light, cool down, and spatter for the different loss functions dice coefficient loss and focal loss, as well as the approach with three different models and one model for all data.

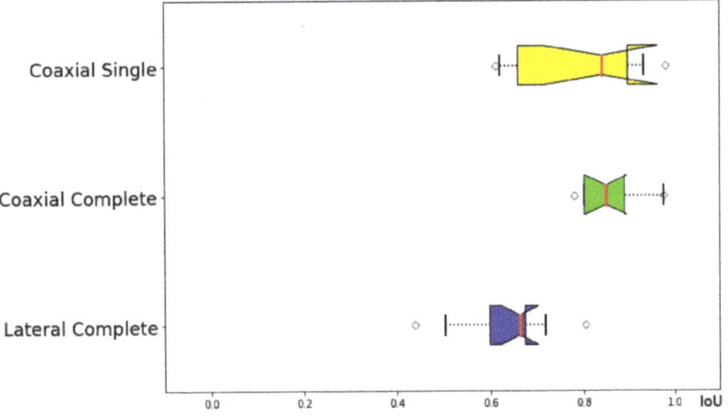

Figure 5. Comparison of the IoU of the different input data sets. All three models are trained with the weighted dice coefficient loss.

Visualized results of the different methods are shown in Table 3. In comparison to the small SDU-Net, results of small U-Net, binary opening, and gray-scale opening are shown. The models of the SDU-Net and the U-Net are trained on all data and with four classes, while the morphological filter on the one hand requires a structural element of different sizes per data set and also cannot distinguish between process light and cooling process. For better visualization, the pixel-by-pixel classification of the neural networks is displayed in different colors and superimposed on the input image. The class of the process light is shown in green color, the spatters in red, and the cooling process in blue. The resulting images of the morphological filter are displayed analogously to Figure 3.

With the morphological filtering, small regions always remain at the edge of the area of the process light, since the structural element can never fill the shape exactly. As a result, the exclusion method would always recognize some pixels as spatter, which must be filtered in post-processing. Even small reflections, which occur mainly in the lateral images, are detected as spatter by the exclusion procedure. On the other hand, spatter that is larger than the defined structural element is detected as process light and not as spatter, which also leads to a wrong result. Compared to the binary opening, the steam generated during welding, which is mainly visible on the lateral images, is usually detected as process light in the gray-scale opening. With the binary opening, the steam area is usually already eliminated during binarization.

Without runtime optimizers, our classification time of the small SDU-Net model on the CPU is about 20 ms. In comparison, the binary and gray-scale opening reached 12 ms for coaxial single, 1.61 ms for coaxial complete, and 40 ms for the lateral complete images. The deviating process times of the opening operation are caused by the different image sizes and the different sizes of the structural element. By using a larger or differently shaped structural element, the process times can be further improved, but the resulting detection quality suffers.

The production time of a stator is about one minute for all 160 to 220 pairs of hairpins. In the best case, 270 ms are needed for welding one hairpin. The quality assurance with the SDU-Net needs 20 ms which is not even 10% of the welding time and 0.33 per mill of the whole welding process. With a time-delayed evaluation of the images, the previous pin can be evaluated each time the next pin is welded. Thus the time sequence of the welding of a stator remains unaffected by this setup due to the fast prediction time. The evaluation was deliberately calculated on the CPU since the model is to be executed directly at the production plant on an industrial PC, where GPUs are not always available. Thus, a strong spatter formation, which can indicate a drifting process or contaminated material, can be reported directly to the user. This allows the user to react directly and stop or adjust the process.

Table 3. Overlaid result images of the different methods. The green color marks the process light, red the spatters, and blue the cooling process. Average IoU value of the pixel which was assigned to process light, spatters, or cooling process.

Image	Ground Truth	SDU-Net	U-Net	Binary Opening	Gray-Scale Opening
Coaxial Single					
		IoU = 0.88	IoU = 0.86	IoU = 0.82	IoU = 0.82
		IoU = 0.79	IoU = 0.522	IoU = 0.482	IoU = 0.471
		IoU = 0.986	IoU = 0	IoU = 0	IoU = 0
Coaxial Complete					
		IoU = 0.79	IoU = 0.79	IoU = 0.71	IoU = 0.70

Table 3. *Cont.*

Image	Ground Truth	SDU-Net	U-Net	Binary Opening	Gray-Scale Opening
		Lateral Complete			
		IoU = 0.68	IoU = 0.65	IoU = 0.37	IoU = 0.37
		IoU = 0.50	IoU = 0.34	IoU = 0.32	IoU = 0.31

4. Discussion and Outlook

By training one model on all three input data sets, it could be shown that a high variance of data can be covered with this approach. The data varies greatly in terms of both recording position and spatter optics. This high level of data variance will not occur in a production line, where a position of data acquisition, as well as a type of pre-processing, will be determined. However, this experiment suggests that it will be possible to use one model for different applications and with slightly different recording mechanisms. We obtain an average IoU of the specific classes without background class of 0.759 in this approach. In comparison, the IoU values of the morphological filter are 0.544 and 0.542, even though these methods were parameterized specifically for the particular data set. This generalization opens the possibility of using one model for different optics without having to make adjustments. With the execution time of 20 ms we are also in a similar range as the execution times of the morphological filter. This requires 12 ms, 1.61 ms, or 40 ms depending on the input data. For some input data, especially *coaxial complete* with 1.6 ms, this process takes less time, but for other input data it takes even longer.

The spatter prediction works on the summed images as well as on the single images. These show the average IoU value for the specific class with 0.821 for *coaxial single*, compared to 0.810 for *coaxial complete*. To record single images, where every spatter is shown punctually, a high recording frequency is required. Cheaper hardware usually has a lower capture frequency, which means that individual spatters would be missed. To counteract this, the exposure time can be increased so that the spatters are visible as lines on the images, similar to the cumulative images we used. The tests showed that even in this application, the spatter can be detected in the image and thus cheaper hardware can be used for quality monitoring.

By using a segmentation approach and a model architecture which works well with strong data augmentation, it is possible to work with very small training data sets. This makes the labeling effort for new processes, e.g., to new customer data, manageable, and thereby saves time and costs. By using a small network architecture with few parameters, both the training time as well as the prediction time are short. Thanks to the short prediction time, the application can be run directly on the production line on a conventional industrial computer. By analyzing the data during production, it is possible to react interactively, which is more efficient than a completely downstream analysis. The algorithm can also be continuously optimized by feeding new data into the neural network under defined monitoring conditions and then training it further. Further knowledge can be generated through the proper application of data feedback. However, in this application, it is important to ensure that the application is not retrained by a drifting process. In addition, an online learning approach for the laser parameters would also be conceivable. The algorithm can be used to check whether spatter occurs with a certain configuration and thus readjust the laser settings.

The data can be recorded coaxially through the laser optics or laterally to the welding process. Spatter detection works well with both recording methods. In the average IoU of the process light and the spatter class, we achieve 0.850 for the coaxial view, while we only achieve 0.688 for the lateral images. It should be noted that the input images in both cases look very different and the relevant image area has different sizes. In the lateral images, a larger area is covered. In both cases, care must be taken to ensure that the distance to the weld seam is large enough to ensure that the spatter is still within the camera's field of view. The coaxial camera setup is often already available on production lines and can therefore be integrated more easily. In this case, the spatter detection could be upgraded in a production line with the help of a software update.

When considering the entire welding process, welding monitoring with a focus on spatter can be seen as just one part of a desired automated 100% inspection of the welding result. This step could be integrated into a three-stage quality monitoring system: in the first step, a deviating position of the hairpins can be detected in the process preparation and thus the welding position can be corrected. In addition, it makes sense to integrate a check of the presence of both hairpins and their correct parallel position. In the second step, spatter monitoring can be carried out directly in the process. This provides information on whether the welding process is unstable and enables rapid response. In the third step, subsequent quality control of the welding results can be carried out. Due to the in-process monitoring, random samples are sufficient in this step.

However, if 100% monitoring for spatter occurrence is to be implemented, additional hardware is required. As mentioned before, an industrial camera installed at production lines usually does not have such a high frame rate that images can be recorded without short times were spatters can be missed. This can be counteracted with the help of an extended exposure time and a larger field of view in which the spatter can be detected, but a 100% view during the process is unrealistic. In this case, an event-based camera or other sensor technology would have to be used. The approach presented in this paper focuses on quick and easy integration into an existing production system without the need for investment in additional hardware. This is often very costly and can lead to additional calibration effort.

Further on, the presented approach can be extended by an additional consideration of laser parameters or other sensor technology, which is already installed in the system. With the help of the information fusion in which the camera-based in-process monitoring for spatter is integrated, it is also possible to control the process even more comprehensively with existing hardware.

Author Contributions: Conceptualization, O.B.; methodology, J.H., A.J. and M.H.; software, J.H.; validation, J.H., A.J. and M.H.; formal analysis, J.H.; data curation, O.B., J.H. and A.J.; writing—original draft preparation, J.H.; writing—review and editing, all authors. All authors have read and agreed to the published version of the manuscript.

Funding: This research received no external funding.

Institutional Review Board Statement: Not applicable.

Informed Consent Statement: Not applicable.

Data Availability Statement: Not applicable.

Conflicts of Interest: The authors declare no conflict of interest.

References

1. Sievi, P.; Käfer, S. Hairpins für E-Mobility Schweißen (German). 2021. Available online: https://www.maschinenmarkt.vogel.de/hairpins-fuer-e-mobility-schweissen-a-1003621/ (accessed on 28 June 2021).
2. Kaliudis, A. It's Heading This Way. 2018. Available online: https://www.trumpf.com/en_INT/presse/online-magazine/its-heading-this-way/ (accessed on 28 June 2021).
3. Kaliudis, A. Green Light for Welding Copper. 2017. Available online: https://www.trumpf.com/en_INT/presse/online-magazine/green-light-for-welding-copper/ (accessed on 28 June 2021).
4. Schmidt, P.A.; Zaeh, M.F. Laser beam welding of electrical contacts of lithium-ion batteries for electric- and hybrid-electric vehicles. *Prod. Eng.* **2015**, *9*, 593–599. [CrossRef]
5. Welding Copper—One Application, a Few Challenges. 2021. Available online: https://www.trumpf.com/en_INT/solutions/applications/laser-welding/welding-copper/ (accessed on 28 June 2021).
6. Dold, E.M.; Willmes, A.; Kaiser, E.; Pricking, S.; Killi, A.; Zaske, S. Qualitativ hochwertige Kupferschweißungen durch grüne Hochleistungsdauerstrichlaser (German). *Metall* **2018**, *72*, 457–459.
7. Franco, D.; Oliveira, J.; Santos, T.G.; Miranda, R. Analysis of copper sheets welded by fiber laser with beam oscillation. *Opt. Laser Technol.* **2021**, *133*, 106563. [CrossRef]
8. Kaplan, A.F.; Powell, J. Laser welding: The spatter map. In Proceedings of the 29th International Congress on Applications of Lasers and Electro-Optics, ICALEO 2010—Congress Proceedings, Anaheim, CA, USA, 26–30 September 2010; Volume 103. [CrossRef]
9. Zhang, M.; Chen, G.; Zhou, Y.; Li, S.; Deng, H. Observation of spatter formation mechanisms in high-power fiber laser welding of thick plate. *Appl. Surf. Sci.* **2013**, *280*, 868–875. [CrossRef]
10. Weigelt, M.; Mayr, A.; Seefried, J.; Heisler, P.; Franke, J. Conceptual design of an intelligent ultrasonic crimping process using machine learning algorithms. *Procedia Manuf.* **2018**, *17*, 78–85. [CrossRef]
11. Mayr, A.; Weigelt, M.; Lindenfels, J.V.; Seefried, J.; Ziegler, M.; Mahr, A.; Urban, N.; Kuhl, A.; Huttel, F.; Franke, J. Electric Motor Production 4.0—Application Potentials of Industry 4.0 Technologies in the Manufacturing of Electric Motors. In Proceedings of the 2018 8th International Electric Drives Production Conference, EDPC 2018—Proceedings, Schweinfurt, Germany, 4–5 December 2018. [CrossRef]
12. Mayr, A.; Lutz, B.; Weigelt, M.; Glabel, T.; Kibkalt, D.; Masuch, M.; Riedel, A.; Franke, J. Evaluation of Machine Learning for Quality Monitoring of Laser Welding Using the Example of the Contacting of Hairpin Windings. In Proceedings of the 2018 8th International Electric Drives Production Conference, EDPC 2018—Proceedings, Schweinfurt, Germany, 4–5 December 2018. [CrossRef]
13. Khumaidi, A.; Yuniarno, E.M.; Purnomo, M.H. Welding defect classification based on convolution neural network (CNN) and Gaussian Kernel. In Proceedings of the 2017 International Seminar on Intelligent Technology and Its Application: Strengthening the Link Between University Research and Industry to Support ASEAN Energy Sector, ISITIA 2017—Proceeding, Surabaya, Indonesia, 28–29 August 2017; Volume 2017-January. [CrossRef]
14. Hartung, J.; Jahn, A.; Stambke, M.; Wehner, O.; Thieringer, R.; Heizmann, M. Camera-based spatter detection in lasser welding with a deep learning approach. In *Forum Bildverarbeitung 2020*; KIT Scientific Publishing: Karlsruhe, Germany, 2020. [CrossRef]
15. Leitz, A. Laserstrahlschweißen von Kupferund Aluminiumwerkstoffen in Mischverbindung (German). Ph.D. Thesis, Universität Stuttgart, Stuttgart, Germany, 2015.
16. Ronneberger, O.; Fischer, P.; Brox, T. U-net: Convolutional networks for biomedical image segmentation. In *Lecture Notes in Computer Science (Including Subseries Lecture Notes in Artificial Intelligence and Lecture Notes in Bioinformatics)*; Springer: Cham, Switzerland, 2015; Volume 9351. [CrossRef]
17. Dosovitskiy, A.; Fischer, P.; Springenberg, J.T.; Riedmiller, M.; Brox, T. Discriminative unsupervised feature learning with exemplar convolutional neural networks. *IEEE Trans. Pattern Anal. Mach. Intell.* **2016**, *38*, 1734–1747. [CrossRef]
18. Tabernik, D.; Šela, S.; Skvarč, J.; Skočaj, D. Segmentation-based deep-learning approach for surface-defect detection. *J. Intell. Manuf.* **2020**, *31*, 759–776. [CrossRef]
19. Redmon, J.; Divvala, S.; Girshick, R.; Farhadi, A. You only look once: Unified, real-time object detection. In Proceedings of the IEEE Conference on Computer Vision and Pattern Recognition, Las Vegas, NV, USA, 27–30 June 2016; Volume 2016-December. [CrossRef]
20. Liu, W.; Anguelov, D.; Erhan, D.; Szegedy, C.; Reed, S.; Fu, C.Y.; Berg, A.C. SSD: Single shot multibox detector. In Proceedings of the European Conference on Computer Vision, Amsterdam, The Netherlands, 11–14 October 2016; Volumee 9905 LNCS. [CrossRef]
21. Wang, S.; Hu, S.Y.; Cheah, E.; Wang, X.; Wang, J.; Chen, L.; Baikpour, M.; Ozturk, A.; Li, Q.; Chou, S.H.; et al. U-Net using stacked dilated convolutions for medical image segmentation. *arXiv* **2020**, arXiv:2004.03466.
22. Zhao, F.; Wu, Z.; Wang, L.; Lin, W.; Gilmore, J.H.; Xia, S.; Shen, D.; Li, G. Spherical Deformable U-Net: Application to Cortical Surface Parcellation and Development Prediction. *IEEE Trans. Med. Imaging* **2021**, *40*, 1217–1228. [CrossRef] [PubMed]
23. Guo, C.; Szemenyei, M.; Pei, Y.; Yi, Y.; Zhou, W. SD-Unet: A Structured Dropout U-Net for Retinal Vessel Segmentation. In Proceedings of the 2019 IEEE 19th International Conference on Bioinformatics and Bioengineering, BIBE 2019, Athens, Greece, 28–30 October 2019. [CrossRef]

24. Gadosey, P.K.; Li, Y.; Agyekum, E.A.; Zhang, T.; Liu, Z.; Yamak, P.T.; Essaf, F. SD-UNET: Stripping down U-net for segmentation of biomedical images on platforms with low computational budgets. *Diagnostics* **2020**, *10*, 110. [CrossRef] [PubMed]
25. Lin, T.Y.; Goyal, P.; Girshick, R.; He, K.; Dollar, P. Focal Loss for Dense Object Detection. *IEEE Trans. Pattern Anal. Mach. Intell.* **2020**, *42*, 2999–3007. [CrossRef] [PubMed]

applied
sciences

Article

In-Process Measurement of Three-Dimensional Deformations Based on Speckle Photography

Andreas Tausendfreund [1,*], Dirk Stöbener [1,2] and Andreas Fischer [1,2]

[1] Bremen Institute for Metrology, Automation and Quality Science, University of Bremen, Linzer Str. 13, 28359 Bremen, Germany; d.stoebener@bimaq.de (D.S.); andreas.fischer@bimaq.de (A.F.)
[2] MAPEX Center for Materials and Processes, University of Bremen, P.O. Box 330440, 28334 Bremen, Germany
[*] Correspondence: a.tausendfreund@bimaq.de or tau@bimaq.de

Abstract: In the concept of the process signature, the relationship between a material load and the modification remaining in the workpiece is used to better understand and optimize manufacturing processes. The basic prerequisite for this is to be able to measure the loads occurring during the machining process in the form of mechanical deformations. Speckle photography is suitable for this in-process measurement task and is already used in a variety of ways for in-plane deformation measurements. The shortcoming of this fast and robust measurement technique based on image correlation techniques is that out-of-plane deformations in the direction of the measurement system cannot be detected and increases the measurement error of in-plane deformations. In this paper, we investigate a method that infers local out-of-plane motions of the workpiece surface from the decorrelation of speckle patterns and is thus able to reconstruct three-dimensional deformation fields. The implementation of the evaluation method enables a fast reconstruction of 3D deformation fields, so that the in-process capability remains given. First measurements in a deep rolling process show that dynamic deformations underneath the die can be captured and demonstrate the suitability of the speckle method for manufacturing process analysis.

Keywords: speckle photography; in-process measurement; deep rolling process

Citation: Tausendfreund, A.; Stöbener, D.; Fischer, A. In-Process Measurement of Three-Dimensional Deformations Based on Speckle Photography. *Appl. Sci.* **2021**, *11*, 4981. https://doi.org/10.3390/app11114981

Academic Editor: Kijoon Lee

Received: 6 May 2021
Accepted: 25 May 2021
Published: 28 May 2021

Publisher's Note: MDPI stays neutral with regard to jurisdictional claims in published maps and institutional affiliations.

Copyright: © 2021 by the authors. Licensee MDPI, Basel, Switzerland. This article is an open access article distributed under the terms and conditions of the Creative Commons Attribution (CC BY) license (https://creativecommons.org/licenses/by/4.0/).

1. Introduction

The functionality of technical surfaces is significantly influenced by the manufacturing process [1–3]. This applies in particular to processes which aim to change the surface layer properties of the workpiece in a suitable manner. However, the targeted generation of, e.g., depth hardness gradients is challenging—even under laboratory conditions [2]. The reasons are that material loads occurring during the machining process in the form of displacements and strains, which are responsible for the material modifications, cannot be measured and that a lack of process knowledge exists. Therefore, the targeted material modifications are only adjusted via the machining parameters or process variables of the machine tools in an iterative way [4–6]. A remedy can be provided by the introduced concept of process signatures [7], which focuses on the internal mechanisms and the influencing quantities [8], leading to a material change to describing the workpiece modification. To gain a deeper understanding of the manufacturing process, the study of process signatures, which correlate the material changes occurring with the acting internal mechanical material loads, is ongoing. The use of process signatures not only enables a better understanding of the process, but also allows a comparison of seemingly different manufacturing processes. As a result, processes and process chains can be optimized to improve functional properties and reduce manufacturing costs [7].

Although a variety of characterization methods for the resulting material modifications exists, the metrological recording of the material loads in the ongoing production process remains a needed prerequisite. Mechanical workpiece loads can be calculated by comparing or correlating images taken before and during the occurrence of the deformation. In contrast to Digital Image Correlation (DIC), the method of Digital Speckle Photography (DSP) [9] is much more precise, especially on low-contrast surfaces. In terms of its maximum achievable resolution, the method is limited only by Heisenberg's uncertainty principle [10]. The in-process capability and suitability of the speckle photography method for two-dimensional deformation measurement in the image plane with resolutions of less than 20 nm [11] has already been demonstrated in several manufacturing applications [12–15]. However, since mechanical machining causes at least a biaxial stress state and, thus, a triaxial strain state in the workpiece via the machining forces, a reconstruction of three-dimensional deformation fields is required for a complete metrological description of the loads.

Even though original speckle photography is only suitable for two-dimensional in-plane measurements, various method derivatives exist to obtain out-of-plane deformation information as well. In this context, Khodadad et al. [16] propose a combination of speckle photography and speckle interferometry. The robust setup uses only one camera and can determine three-dimensional displacements with a resolution of 0.6 µm. The method introduced by Fricke-Begemann [17] uses the decorrelation of the speckle patterns associated with the out-of-plane motion, i.e., the speckle correlation functions changing with the motion are evaluated. The determined measurement uncertainties here are significantly lower than in the approach of Khodadad et al. [16] and are in the range of one tenth of the wavelength.

Therefore, the aim of this paper is the realization of a solely speckle-photographic in-process measurement system for the acquisition of three-dimensional load fields. In a first practical implementation, the deformation fields in a manufacturing rolling process are to be measured. The open questions in the context of speckle-photographic 3D deformation measurement in manufacturing processes are to what extent such a measurement system is suitable for use in the harsh manufacturing environment and to what extent the proposed algorithms are sufficient for in-process measurement. Furthermore, it must be clarified in which functional relationship the speckle decorrelation is related to displacements in the out-of-plane axis and how a calibration of this axis needs to be performed.

Accordingly, this paper is organized as follows: Section 2 describes the challenges of three-dimensional deformation measurement by speckle photography, especially in deformation evaluation according to Fricke-Begemann [17]. Moreover, an alternative approach and its implementation in the context of sub-pixel interpolation is presented. After the demonstration of calibration strategies, the first in-process measurements during deep rolling follow in Section 3. Section 4 discusses the reconstruction of the three-dimensional deformation or displacement field, followed by a summary and an outlook in Section 5.

2. Materials and Methods

2.1. In-Plane Speckle Photography

When illuminating a rough surface, a so-called speckle pattern is formed in the image plane of an imaging camera system by constructive and destructive interference (see zoomed area W_{eval} in Figure 1). Here, each surface point on the object surface, with the coordinates ζ and ψ, is assigned to a point (x,y) on the camera chip. If a point (ζ,ψ) of the surface is shifted in-plane by $\Delta\zeta$ or $\Delta\psi$ due to deformation, the corresponding point (x,y) in the image plane is shifted by Δx or Δy, respectively. This shift can be analyzed via the maximum of the cross-correlation function according to Figure 1. In order to measure a global deformation field of many object points in a region of interest W_{FOV}, a small local evaluation window W_{eval} is rasterized across the overall image W_{FOV}.

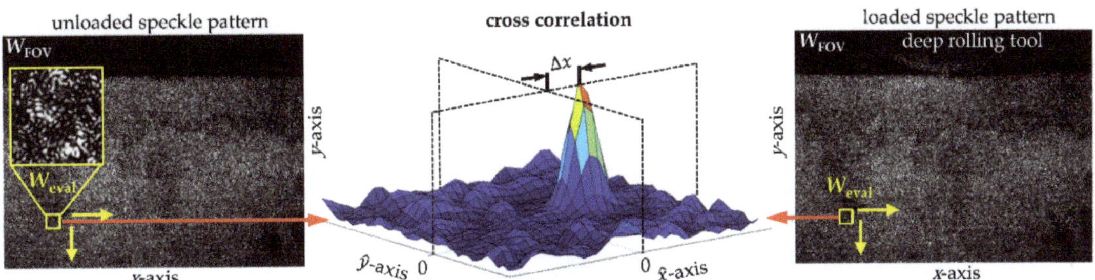

Figure 1. Evaluation principle of speckle photography for the measurement of the two-dimensional in-plane deformation.

In order to determine the position of the maximum of the cross-correlation that is shown in Figure 1 with sub-pixel accuracy, interpolation methods are used based on curve approximation. The currently used sub-pixel interpolation algorithm, according to Nobach and Honkanen [18], determines the shift of the maximum $(\Delta x, \Delta y)$ using an elliptic Gaussian function

$$g(\hat{x}, \hat{y}) = a \cdot \exp\left\{ b_{20}(\hat{x} - \Delta x)^2 + b_{11}(\hat{x} - \Delta x)(\hat{y} - \Delta y) + b_{02}(\hat{y} - \Delta y)^2 \right\}, \tag{1}$$

with the variables $(\hat{x}, \hat{y})$, which is fitted into to the cross-correlation function. It can be rewritten as:

$$g(\hat{x}, \hat{y}) = \exp\left\{ c_{00} + c_{10}\hat{x} + c_{20}\hat{x}^2 + c_{01}\hat{y} + c_{11}\hat{x}\hat{y} + c_{02}\hat{y}^2 \right\} \tag{2}$$

with

$$c_{00} = \ln(a) + b_{20}\Delta^2 x + b_{11}\Delta x \Delta y + b_{02}\Delta^2 y, \tag{3}$$

$$c_{10} = -2b_{20}\Delta x - b_{11}\Delta y, \tag{4}$$

$$c_{20} = b_{20}, \tag{5}$$

$$c_{01} = -b_{11}\Delta x - 2b_{02}\Delta y, \tag{6}$$

$$c_{11} = b_{11}, \tag{7}$$

$$c_{02} = b_{02}. \tag{8}$$

The shift components Δx, Δy of the maximum are then obtained from the Equations (3)–(8) as follows [18]:

$$\Delta x = \frac{c_{11}c_{01} - 2c_{10}c_{02}}{4c_{20}c_{02} - c_{11}^2}, \tag{9}$$

$$\Delta y = \frac{c_{11}c_{10} - 2c_{01}c_{20}}{4c_{20}c_{02} - c_{11}^2}. \tag{10}$$

The relationship between the displacements Δx, Δy in the image plane and the respective object displacements $\Delta\xi$, $\Delta\psi$ within the measuring area are finally obtained according to the magnification A of the imaging system:

$$\Delta\xi = A\,\Delta x, \tag{11}$$

$$\Delta\psi = A\,\Delta y. \tag{12}$$

2.2. Three-Dimensional Speckle Photography in Out-of-Plane Direction

For the initially targeted objective of an extended understanding of the manufacturing process, the acquisition of information about the 3D deformation field is essential. Speckle photographic approaches are in principle also suitable for this purpose, since the recorded speckle patterns basically represent sections through the three-dimensional structure of the wave field. Figure 2a shows an example of the measured three-dimensional structure of the speckle pattern, for a ζ-shift of the measurement plane.

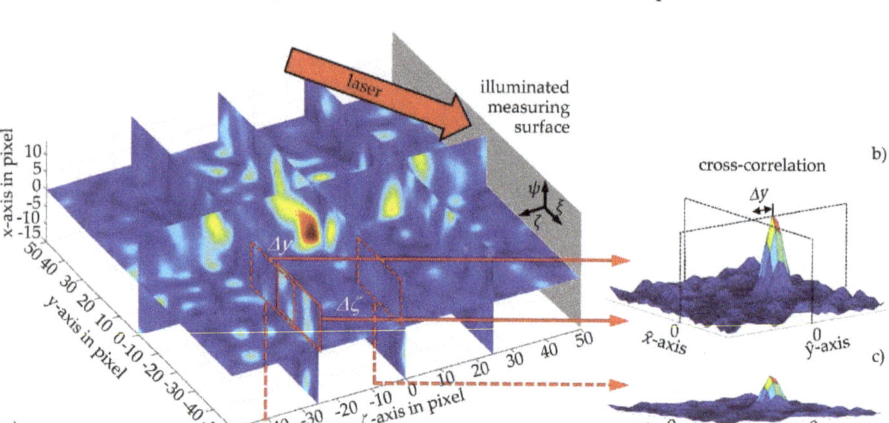

Figure 2. Three-dimensional speckle pattern measured by shifting the camera and illumination system in the ζ-axis (**a**) and the calculated cross-correlation function of two evaluation windows shifted against each other in the *y*-direction (**b**) or in the ζ-direction (**c**).

The cross-correlation for pure in-plane displacements $\Delta\xi$, $\Delta\psi$ is shown in Figure 2b. Deformations and displacements of the surface in the out-of-plane direction ($\Delta\zeta$), however, lead to a decorrelation of the speckle patterns, which means a reduced amplitude of the cross-correlation maximum (compare Figure 2b,c). This decorrelation was previously considered as a disturbance of the measurement effect and was minimized by using higher measurement rates. The approach of Fricke-Begemann uses the decorrelation and derives the out-of-plane deformation of the measurement surface via an elaborate shape analysis of the power spectrum of the cross-correlated speckle patterns [17]. A maximum likelihood estimation is first used to determine the tilt angles of the surface points and based on this, comprehensive three-dimensional deformation measurements are finally reconstructed. However, laboratory investigations of simultaneous tilts and in-plane displacements of the measurement surface illustrate the problem that arises in separating tilt and in-plane displacement in this form of evaluation, see Figure 3.

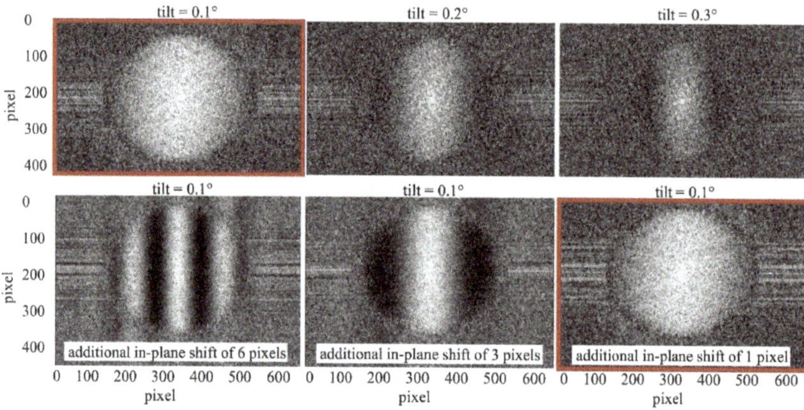

Figure 3. Power spectra of two speckle patterns for tilted or additional displaced measurement surfaces.

Especially in the range of very small in-plane displacements (<1 pixel), as they are predominantly present in the in-process load measurements of manufacturing processes, the differences are hardly recognizable. While in-plane displacements in the cross-correlation (see Figure 1) are clearly recognizable by the displacement of the maximum without significantly influencing the shape of the curve, in the evaluation via the power spectrum both the in-plane displacement and the out-of-plane movement affect the shape of the curve. Even though the power spectrum is very sensitive to the tilts, complex fit and filter algorithms are required to separate the two components. First approaches for the analysis of a two-axis motion, consisting of a tilt in the ξ-axis and a displacement in ψ-direction, resulted in computation times in the double-digit minute range for the smallest evaluation windows. With regard to the parallelization of the evaluation algorithms and the goal of a fast in-process measurement, an alternative approach is therefore pursued. The cross-correlation allows for separating the in-plane and out-of-plane displacement information by evaluating the shift of the correlation maximum (see Figure 2b) and the broadening of the normalized cross-correlation function (see Figure 2c), respectively. The shape or the width of the correlation function is already evaluated in the context of the performed sub-pixel interpolation. Inherently hidden in the time-optimized interpolation is the width σ of the approximated Gaussian function $g(\hat{x}, \hat{y})$. It can be considered as a value for the speckle decorrelation and thus for the out-of-plane deformation. According to the Equations (1), (5) and (8), the width components for the $\hat{x}$- or $\hat{y}$-direction read

$$\sigma_x = \sqrt{\frac{1}{-2c_{20}}}, \tag{13}$$

$$\sigma_y = \sqrt{\frac{1}{-2c_{02}}}. \tag{14}$$

While the functional relationship between the displacement of the maximum of the cross-correlation function and the in-plane displacement of a surface point is linearly related via the imaging scale A, the functional relationship between the out-of-plane shift $\Delta\zeta$ and the width of the correlation function σ_x or σ_y is unknown yet. For the establishment of a respective calibration function, the relation $\Delta\zeta(\sigma_x,\sigma_y)$ is investigated experimentally in the following.

2.3. Relationship between the Width of the Correlation Function and the Out-of-Plane Shift

In order to analyze the functional relationship $\Delta\zeta(\sigma_x,\sigma_y)$, the laboratory setup of a speckle-photographic measurement system was realized, see Figure 4a.

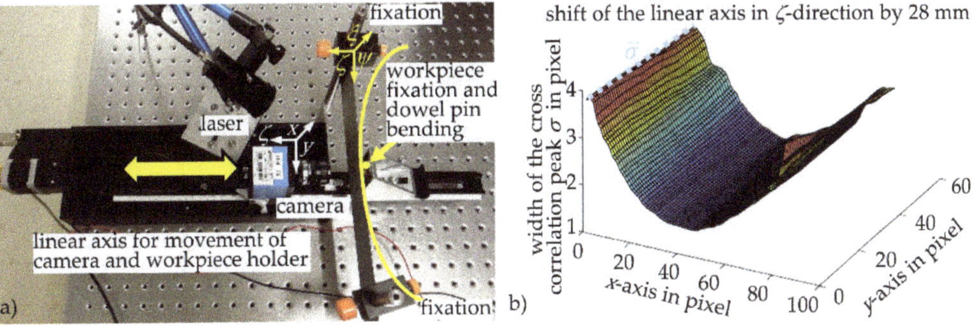

Figure 4. Speckle-photographic laboratory experiment: (**a**) laboratory setup with indicated bending of the workpiece sheet over a locating dowel pin (yellow) fixed in the camera system; (**b**) field of the respective local widths of the correlation functions, at a linear axis displacement of 28 mm.

The measurement object, a sandblasted sheet metal strip (400 mm × 50 mm) is illuminated with a collimated widened Omicron laser beam (λ = 405 nm). The illuminated area is observed and imaged with an Imaging Source camera (DMK 21AU04) through a Pentax 25 mm lens. Camera, laser and the center of the sheet metal strip (in the center of the measurement field) are fixed on a linear axis from PI company, Germany. The outer ends of the sheet metal strip, on the other hand, are fixed to the laboratory bench. Thus, a movement of the linear axis causes a bending of the sheet metal strip (yellow line in Figure 4a) and thus a relative movement between the center and its outer ends.

Figure 4b shows the global field of the respective locally calculated correlation function widths

$$\sigma = \sqrt{\sigma_x^2 + \sigma_y^2} = \sqrt{\frac{1}{-2c_{20}} + \frac{1}{-2c_{02}}} \tag{15}$$

for a linear axis displacement of 28 mm in ζ-direction with a total measurement field size of 30 mm × 40 mm. The parabolic bending around the dowel pin can be seen very well in the center of the image. The width of the correlation function increases significantly from the original 1.3 pixels to about 4 pixels. If looking in each case at the mean value of the width of the cross-correlation function along the y-axis at the position $x = 0$ in Figure 4b, one can see the linear relationship between the respective adjusted displacement of the linear axis and the broadening of the interpolated correlation function $\bar{\sigma}$ shown in Figure 5. The functional relationship between a changed loading condition with a resulting displacement $\Delta\zeta$ and the resulting speckle decorrelation or change in the width $\Delta\sigma$ of the Gaussian fit g from Equation (1) is thus linear in nature, i.e.,

$$\Delta\zeta = \pm B\,\Delta\sigma. \tag{16}$$

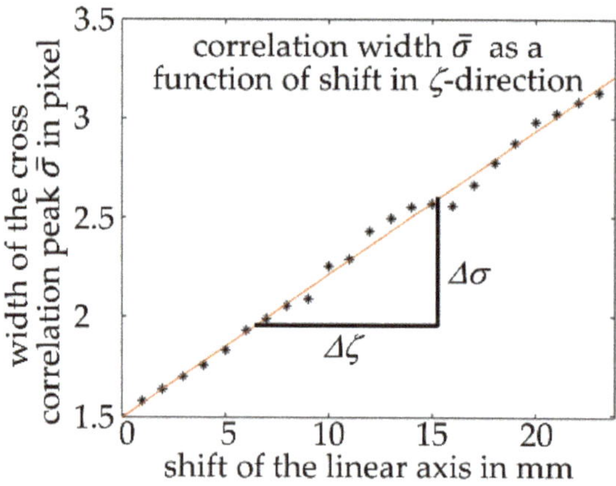

Figure 5. Linear relationship between the width of the correlation function $\bar{\sigma}$ calculated in the sub-pixel interpolation and the shift of the linear axis in ζ-direction.

The indefinite sign for the calibration factor B results intuitively from Figure 2. The shift by $+\Delta\zeta$ leads to the decorrelation of the speckle pattern to the same extent as a shift by $-\Delta\zeta$. While the calibration factor can be defined prior to a respective measurement on the basis of a defined displacement, the sign can only be determined via prior knowledge of the observed deformation process. In the case of the bending test, for example, the outer ends of the sheet metal strip will move away from the camera in negative ζ-direction, so that negative signs are to be expected. While the in-plane deformation $(\Delta\xi,\Delta\psi)$ is linearly related to the displacement $(\Delta x,\Delta y)$ of the maximum of the cross-correlation via the scaling factor A (Equations (11) and (12)), it is shown by Equation (16) that a linear relationship also results for the out-of-plane deformation $\Delta\zeta$ via the factor B and the change in width of the cross-correlation function $\Delta\sigma$.

3. Results of In-Process Measurements in a Deep Rolling Process

3.1. Measurement Setup and Experimental Realization

Due to the expected strong out-of-plane movement in the edge area of the machined surface, the deep rolling process was selected as the manufacturing process to investigate first. The realized manufacturing process with the corresponding measurement setup is shown in Figure 6. Contrary to the intuitive idea, the setup is upside down and the rolling tool presses on the workpiece from below. The design was therefore chosen so that any cooling lubricant escaping from the hydraulic rolling tool is prevented by gravity from contaminating the measuring field located on the front surface and disturbing the measurement. The rotation axis of the cylindrical tungsten carbide tool with a diameter of 13 mm, which protrudes a few millimeters above the workpiece, is located in the ζ-axis of the specimen or measuring system. The $\zeta\psi$-plane lies in the face of the hardened martensitic 42CrMo4 workpiece (150 mm $\times$ 70 mm $\times$ 20 mm). The tool moves while rolling with a loading of 350 bar in the negative ζ-direction with a feed rate of v = 0.02 mm/s (see Figure 6a).

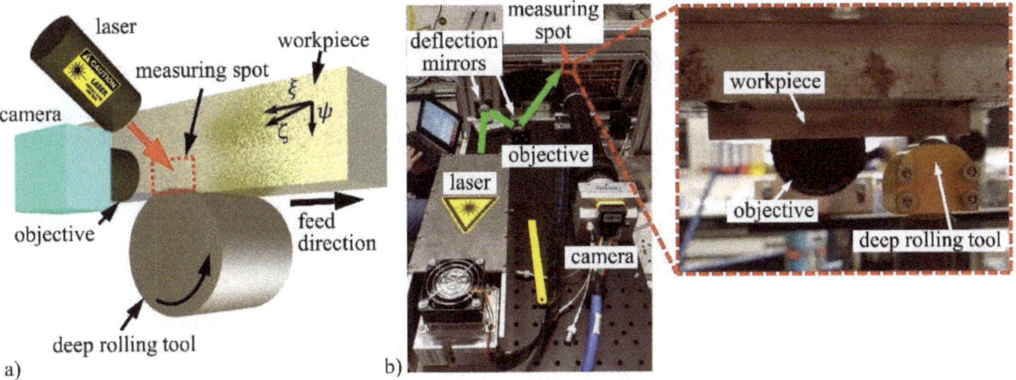

Figure 6. Test setup for deep rolling: Basic measurement setup (**a**) and practical implementation with enlarged view from the opposite ζ-direction to the camera (**b**).

The 8 mm $\times$ 6 mm measuring field is illuminated by a short pulse laser (t_p = 1 ns, λ = 532 nm) from Horus via two deflection mirrors and observed with a 4K high speed camera CP70 from Optronis via a 90 mm Apo-Rodagon lens from Rodenstock (see Figure 6b). In the rolling process with its moderate feed rate, full-frame recordings could be made at a recording rate of 166 frames per second. The implemented manufacturing setup is also used for in situ X-ray diffraction analysis and residual stress measurement and is described in detail in [19].

3.2. Results of 3D-Deformation Measurements

The results of a three-dimensional load measurement for the deep rolling test are shown in Figure 7. Assuming a homogeneous workpiece, the evaluation range in ζ-direction is extended from 8 mm to 20 mm. For this purpose, the speckle-photographic deformation measurements of the time series—in the direction of movement of the tool—were stitched into an extended evaluation field [15,20]. The calibration of the ζ- and ψ-axis is done according to Equations (11) and (12) from the magnification A of the lens system.

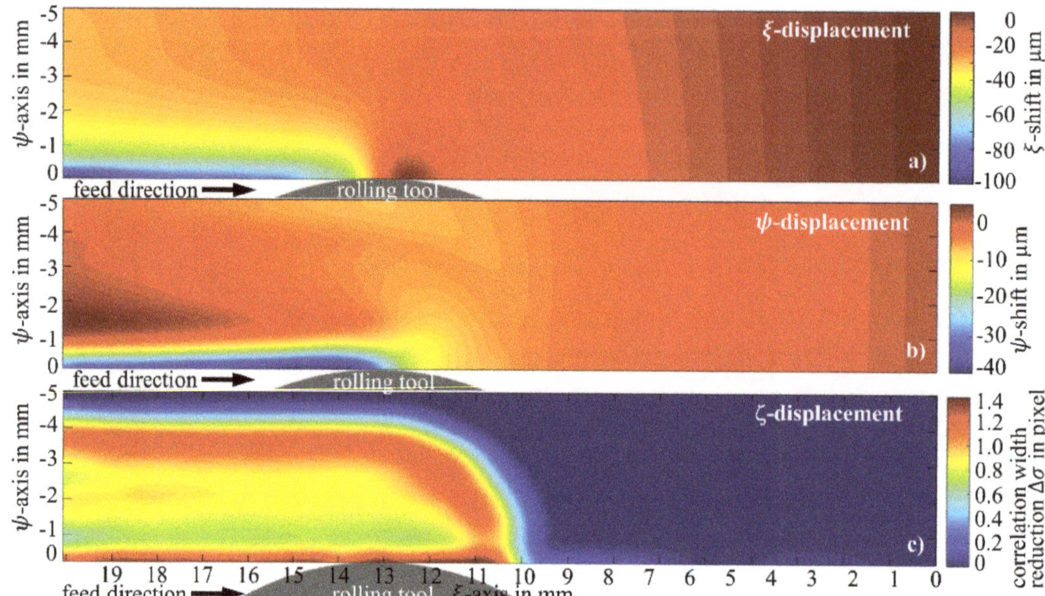

Figure 7. In-process measurement in the deep rolling process: The occurring shifts in ξ-direction (**a**) and Ψ-direction (**b**) as well as the reduction of the calculated width of the correlation function resulting from the out-of-plane shift as a value for the ζ-shift (**c**) are shown.

The load measurements in the running deep rolling process show maximum deformation differences for the ξ- and ψ-direction at a depth of 800 μm. Accordingly, the equivalent stress maximum also occurs at this depth, which is in line with the expectations [19]. At the machined surface, the material is moved up to 100 μm in the direction of the tool movement (see Figure 7a). The minimum displacement in ψ-direction is located about 1 mm behind the contact point of the tool at position 14 mm and is -40 μm (see Figure 7b). In addition to the measured in-plane deformations in ξ- and ψ-direction, Figure 7c shows for the first time the out-of-plane component of the in-process characterization in the direction of the camera axis. The determined deformation in ζ-direction corresponds to the broadening of the correlation function or a decorrelation of the speckle patterns between unloaded and loaded state.

4. Discussion

For a complete description of the three-dimensional displacement field, the only remaining task is to perform the calibration between the $\Delta\zeta$-displacement and the decrease of the correlation width $\Delta\sigma$ according to Equation (16). In this case, the scaling or calibration of the ζ-axis is performed by a subsequent reference measurement with a stylus instrument (Mahr LD-120). For this purpose, the tactilely measured burr (see Figure 8a) was compared with the determined width of the correlation function (see Figure 8b). The approximately linear deformation of the burr in the section shown is reflected proportionally in the linear behavior of the correlation width $\Delta\sigma$ and confirms the preliminary tests from Figure 5. Therefore, a linear correlation is identified and the ζ-axis is accordingly scaled for the speckle photographic measurement in Figure 8c. The evaluation of the speckle decorrelation basically does not provide any information about the sign of the measured $\Delta\zeta$-deformation (see Section 2.3). However, the evaluation succeeds due to the prior knowledge gained from the calibration measurements that the material is pressed outward in the direction of the camera.

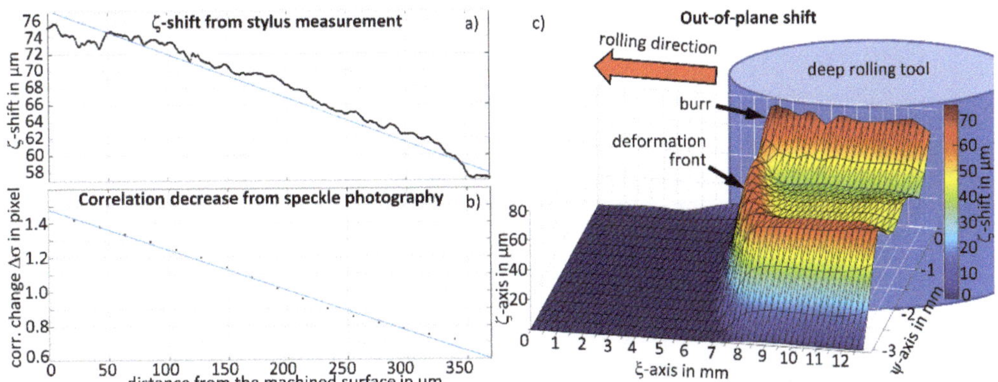

Figure 8. 3D deformation measurement in the deep rolling process. (**a**) Tactilely measured burr; (**b**) width of the correlation function as a function of the out-of-plane displacement of the burr; (**c**) speckle-photographically measured deformation field in the ζ-direction after calibration.

Figure 8c clearly shows the strong burr formation at the edge of the workpiece with an extension of up to 80 µm, caused by the high rolling pressure of 350 bar applied to the 42CrMo4 sample. Highlighted in the measured data is the dynamic deformation front, which propagates like a 'bow wave' in front of the tool. It is exactly this dynamic deformation front that has so far caused problems in the interpretation of the measured data, since the accompanying decorrelation of the speckle patterns could not be explained. The acquisition of load dynamics shows that speckle photography is suitable for measuring and evaluating three-dimensional deformation fields within ongoing manufacturing processes. In particular, the analysis of the load dynamics is a key to the establishment of process signatures described above as well as for a comparison with corresponding simulations of the manufacturing process and illustrates the great potential of this in-process measurement technique.

5. Conclusions and Outlook

Laboratory investigations in the context of the analysis of the functional relationship between the decorrelation of measured speckle patterns or the broadening of the correlation function $\Delta\sigma$ and the out-of-plane deformation $\Delta\zeta$ confirm a linear behavior. Moreover, the additional computational effort for the determination of $\Delta\sigma$ in the sub-pixel evaluation is negligible, which means that the parallelizability of the algorithms remains given and three-dimensional in-process deformation measurements during the rolling process become possible. The measurements during deep rolling show the dynamic deformations below the tool. From these deformations, the strain and thus the load during the die engagement can be calculated directly via the formation of the gradients. In particular, the 'bow wave' of out-of-plane deformation moving ahead of the tool shows that the local mechanical material loads can be much more pronounced than the workpiece modifications remaining in the material. In order to better understand manufacturing processes, 3D speckle photography can thus make an important contribution in the context of the sought-after process signatures.

While the calibration of the ζ- and ψ-displacement can be performed via the imaging scale of the camera system, this is only possible for the ζ-axis via a subsequent reference measurement with a stylus instrument. The ambiguity with regard to the direction of the displacement can also be eliminated by the additional information from the reference measurement. In order to ensure in-process capability in various manufacturing processes, the realization of a self-calibration of the measuring system is planned for the future. For this purpose, a three-axis piezo stack is to be used, via which the system can be calibrated before each individual measurement by means of defined, homogeneous displacements. In addition to the calibration, it is thus simultaneously possible to eliminate optical distortions—for example due to misalignments of the measuring system—already during the measurement and to carry out quantitative 3D deformation measurements in running manufacturing processes.

Author Contributions: Conceptualization, A.T.; investigation, A.T.; methodology, D.S. and A.F.; supervision, A.F.; writing—original draft, A.T.; writing—review and editing, D.S. and A.F. All authors have read and agreed to the published version of the manuscript.

Funding: This research was funded by the German Research Foundation (DFG) for subproject C06 'Surface optical measurement of mechanical working material loads' within the Transregional Cooperative Research Center SFB/TRR136.

Institutional Review Board Statement: Not applicable.

Informed Consent Statement: Not applicable.

Data Availability Statement: The data presented in this study are not available.

Acknowledgments: The authors thank Heiner Meyer (SFB/TRR136, subproject C01) and their colleagues from the Leibniz Institute for Materials Engineering IWT for the friendly support during the conduction of the in-process measurements.

Conflicts of Interest: The authors declare no conflict of interest. The founding sponsors had no role in the design of the study; in the collection, analyses, or interpretation of data; in the writing of the manuscript, and in the decision to publish the results.

References

1. Field, M.; Kahles, J.F. Review of Surface Integrity of Machined Components. *CIRP Ann. Manuf. Technol.* **1971**, *20*, 153–163.
2. Jawahir, I.S.; Brinksmeier, E.; M'Saoubi, R.; Aspinwall, D.K.; Outeiro, J.C.; Meyer, D.; Umbrello, D.; Jayal, A.D. Surface integrity in material removal processes: Recent advances. *CIRP Ann.* **2011**, *60*, 603–626. [CrossRef]
3. Habschied, M.; de Graaff, B.; Klumpp, A.; Schulze, V. Fertigung und Eigenspannungen. *HTM J. Heat Treat. Mater.* **2015**, *70*, 111–121. [CrossRef]
4. Liang, S.Y.; Su, J.-C. Residual Stress Modeling in Orthogonal Machining. *CIRP Ann.* **2007**, *56*, 65–68. [CrossRef]
5. Lazoglu, I.; Ulutan, D.; Alaca, B.E. An enhanced analytical model for residual stress prediction in machining. *CIRP Ann.* **2008**, *57*, 81–84. [CrossRef]
6. Brinksmeier, E.; Heinzel, C.; Garbrecht, M.; Sölter, J.; Reucher, G. Residual Stresses in High Speed Turning of Thin-Walled Cylindrical Workpieces. *Int. J. Autom. Technol.* **2011**, *5*, 313–319. [CrossRef]
7. Brinksmeier, E.; Klocke, F.; Lucca, D.A.; Sölter, J.; Meyer, D. Process Signatures—A New Approach to Solve the Inverse Surface Integrity Problem in Machining Processes. *Procedia CIRP* **2014**, *13*, 429–434. [CrossRef]
8. Greenwood, G.W.; Johnson, R.H. The deformation of metals under small stresses during phase transformations. *Proc. R. Soc. London. Ser. A Math. Phys. Sci.* **1965**, *283*, 403–422.
9. Parks, V.J. The range of speckle metrology. *Exp. Mech.* **1980**, *20*, 181–191. [CrossRef]
10. Fischer, A. Fundamental uncertainty limit of optical flow velocimetry according to Heisenberg's uncertainty principle. *Appl. Opt.* **2016**, *55*, 8787–8795. [CrossRef] [PubMed]
11. Tausendfreund, A.; Stöbener, D.; Ströbel, G. In-process measurements of strain fields during grinding. In Proceedings of the 16th International Conference of the European Society for Precision Engineering and Nanotechnology (Euspen), Nottingham, UK, 30 May–3 June 2016; pp. 85–86.
12. Tausendfreund, A.; Borchers, F.; Kohls, E.; Kuschel, S.; Stöbener, D.; Heinzel, C.; Fischer, A. Investigations on Material Loads during Grinding by Speckle Photography. *J. Manuf. Mater. Process.* **2018**, *2*, 71. [CrossRef]
13. Tausendfreund, A.; Stöbener, D.; Fischer, A. Precise In-Process Strain Measurements for the Investigation of Surface Modification Mechanisms. *J. Manuf. Mater. Process.* **2018**, *2*, 9. [CrossRef]
14. Tausendfreund, A.; Stöbener, D.; Fischer, A. Induction of Highly Dynamic Shock Waves in Machining Processes with Multiple Loads and Short Tool Impacts. *Appl. Sci.* **2019**, *9*, 2293. [CrossRef]
15. Tausendfreund, A.; Stöbener, D.; Fischer, A. In-process workpiece deformation measurements under the rough environments of manufacturing technology. *Procedia CIRP* **2020**, *87*, 409–414. [CrossRef]
16. Khodadad, D.; Singh, A.K.; Pedrini, G.; Sjödahl, M. Full-field 3D deformation measurement: Comparison between speckle phase and displacement evaluation. *Appl. Opt.* **2016**, *55*, 7735–7743. [CrossRef] [PubMed]
17. Fricke-Begemann, T. Optical Measurement of Deformation Fields and Surface Processes with Digital Speckle Correlation. Ph.D. Thesis, University of Oldenburg, Oldenburg, Germany, 2002.

18. Nobach, H.; Honkanen, M. Two-dimensional Gaussian regression for sub-pixel displacement estimation in particle image velocimetry or particle position estimation in particle tracking velocimetry. *Exp. Fluids* **2005**, *38*, 511–515. [CrossRef]
19. Meyer, H.; Epp, J. In Situ X-ray Diffraction Analysis of Stresses during Deep Rolling of Steel. *Quantum Beam Sci.* **2018**, *2*, 11. [CrossRef]
20. Tausendfreund, A.; Stöbener, D.; Fischer, A. Messung thermomechanischer Beanspruchungen in laufenden Schleifprozessen. *tm-Tech. Mess.* **2020**, *87*, 201–209. [CrossRef]

*applied
sciences*

MDPI

Article

Diffusional Behavior of New Insulating Gas Mixtures as Alternatives to the SF_6-Use in Medium Voltage Switchgear

Ane Espinazo [1,*], José Ignacio Lombraña [1], Estibaliz Asua [2], Beñat Pereda-Ayo [1], María Luz Alonso [3], Rosa María Alonso [3], Leire Cayero [1], Jesús Izcara [4] and Josu Izagirre [4]

[1] Chemical Engineering Department, Faculty of Science and Technology, University of the Basque Country (UPV/EHU), Barrio Sarriena s/n, 48940 Leioa, Spain; ji.lombrana@ehu.eus (J.I.L.); benat.pereda@ehu.eus (B.P.-A.); leirecayerogaray@gmail.com (L.C.)

[2] Electricity and Electronics Department, Faculty of Science and Technology, University of the Basque Country (UPV/EHU), Barrio Sarriena s/n, 48940 Leioa, Spain; estibaliz.asua@ehu.eus

[3] Analytical Chemistry Department, Faculty of Science and Technology, University of the Basque Country (UPV/EHU), Barrio Sarriena s/n, 48940 Leioa, Spain; marialuz.alonso@ehu.eus (M.L.A.); rosamaria.alonso@ehu.eus (R.M.A.)

[4] Ormazabal Corporate Technology, Parque Empresarial Boroa, Parcela 24, 48340 Amorebieta-Etxano, Spain; jih@ormazabal.com (J.I.); josu.izagirre@ormazabal.com (J.I.)

* Correspondence: ane.espinazo@ehu.eus

Citation: Espinazo, A.; Lombraña, J.I.; Asua, E.; Pereda-Ayo, B.; Alonso, M.L.; Alonso, R.M.; Cayero, L.; Izcara, J.; Izagirre, J. Diffusional Behavior of New Insulating Gas Mixtures as Alternatives to the SF_6-Use in Medium Voltage Switchgear. *Appl. Sci.* **2022**, *12*, 1436. https://doi.org/10.3390/app12031436

Academic Editor: Andreas Fischer

Received: 27 December 2021
Accepted: 27 January 2022
Published: 28 January 2022

Publisher's Note: MDPI stays neutral with regard to jurisdictional claims in published maps and institutional affiliations.

Copyright: © 2022 by the authors. Licensee MDPI, Basel, Switzerland. This article is an open access article distributed under the terms and conditions of the Creative Commons Attribution (CC BY) license (https://creativecommons.org/licenses/by/4.0/).

Abstract: Regarding the use of SF_6 in medium voltage switchgear (MVS), a review of alternatives was encouraged by the European Parliament in Regulation No 517/2014. This is aimed at a new regulatory change, that is expected soon, which will include its prohibition, similar to what has happened with other fluorinated greenhouse gases in other fields, like refrigeration. Therefore, there is an urgent need to study the physical and chemical properties of alternative gas mixtures to determine if they are suitable to replace SF_6. In this context, this work addresses the difusional analysis of new gases. Binary and ternary mixtures made of 1,3,3,3-tetrafluoropropene ($C_3F_4H_2$) and heptafluoroisopropyl trifluoromethyl ketone ($C_5F_{10}O$), using dry air as a carrier gas, were studied. The mixtures were analyzed using original equipment, composed of UV-Vis spectroscopy technology in a sealed gas chamber, which is similar to MVS. Consequently, an experimental equipment that monitors the concentration of a gas mixture online and a model that predicts the mixing process were designed and tested. The concentration profiles were obtained concerning both the time and position in the gas chamber, and the diffusional and convectional parameters were numerically calculated and optimized in an algorithm created in Scilab.

Keywords: medium voltage switchgear; SF_6 alternatives; UV-Vis spectroscopy; gas mixing modeling; multicomponent diffusion analysis

1. Introduction

Sulfur hexafluoride, SF_6, is the most widely applied gas in the electric market for insulation and electric arc quenching due to its high stability, high dielectric strength, and non-toxicity [1]. It is commonly used in gas-insulated switchgear (GIS), gas-insulated transformers (GIT), gas-insulated lines (GIL), and gas-insulated circuit breakers (GICB) [2]. However, it is also considered a very strong greenhouse gas, with a global warming potential (*GWP*) of about 23,500 on a 100-year horizon, making SF_6 the compound with the highest value according to the Fifth Assessment Report (AR5) of the Intergovernmental Panel on Climate Change (IPCC) [3]. In 1997, in the Kyoto Protocol [4], the reduction of greenhouse gas emissions was established, which included SF_6, so alternative gases for applications that use SF_6 have been investigated ever since. Although the equipment in which sulfur hexafluoride is used is sealed, leaks between the sealing surfaces, permeation of the gas through the thermoplastic materials, or accidents can happen [5], resulting in gas emissions. SF_6 and other common fluorinated greenhouse gases, their lifetime (*LT*) in the atmosphere, *GWP* values, and applications are shown in Table 1.

Table 1. The lifetime (*LT*) in the atmosphere, global warming potential values in a 100-year horizon (GWP_{100}), and applications of SF_6 and other fluorinated greenhouse gases.

Component	LT (yr) *	GWP_{100} *	Application
Sulfur hexafluoride SF_6	3200	23,500	Electrical insulation, polycrystalline silicone layer etching, blanketing gas for aluminum or magnesium [1]
Nitrogen trifluoride NF_3	500	16,100	Microelectronic equipment cleaning, etching in microelectronics, liquid crystal displays, photovoltaic cells [6]
Difluoromethane CH_2F_2	5.2	677	Refrigeration [7]
Perfluoromethane CF_4	50,000	6630	Semiconductor manufacturing, microelectronic equipment cleaning, etching in microelectronics, chemical vapor decomposition [8]
1,1,1,2-tetrafluoroethane CF_3CHF_2	28.2	3170	Refrigeration, propellant applications, vapor-phase solvent, electronic degreasing [9]
Perfluoroethane C_2F_6	10,000	11,100	Semiconductor manufacturing, microelectronic equipment cleaning, etching in microelectronics, chemical vapor decomposition [8]
1,1,1,3,3-pentafluoropropane $CF_3CH_2CHF_2$	7.7	858	Chemical blowing agent, refrigerant [10]
Perfluoropropane C_3F_8	2600	8900	Refrigeration, semiconductor manufacturing, microelectronic equipment cleaning, etching in microelectronics, chemical vapor decomposition [8]

* Data retrieved from the Fifth Assessment Report (AR5) of the Intergovernmental Panel on Climate Change (IPCC) [3].

In 2014, the European Parliament developed Regulation (EU) No 517/2014 [11], on fluorinated greenhouse gases, where rules for the use, containment, recovery, and destruction of some fluorinated greenhouse gases are collected. These fluorinated gases include not only SF_6, but also hydrofluorocarbons (HFC) and perfluorocarbons (PFC). The European Commission is currently reviewing the impact that the actual regulation has had, and is expected to propose a new regulation soon [12]. For that reason, it is urgent to find an alternative that replaces SF_6 in medium voltage switchgear (MVS) before a possible ban on its use comes into force.

Initially, SF_6 alternatives were focused on the use of existing technologies such as air-insulated switchgear (AIS) [13] or solid-insulated switchgear (SIS) [14]. Even though both have been commonly used in low-voltage equipment, their use in higher voltage systems would require higher dimensions and, therefore, higher costs.

The mixture between SF_6 with a carrier gas, like air, N_2, or CO_2, was then considered for insulation [15]. Although the use of SF_6 is reduced, the environmental impact of the mixture, even with low SF_6 concentrations, is still significantly high. Consequently, researchers are more focused on finding other gas or gas mixtures that replace SF_6 completely.

There are some important requirements that this gas or gas mixture must meet, according to Kieffel et al. [16], to be considered a suitable option: high dielectric strength, good arc quenching capability, low boiling point, high vapor pressure at low temperature, high heat dissipation, and compatibility with the materials used in electrical switchgear, among others. Moreover, the gases must have low toxicity, no flammability, no ozone depletion potential (*ODP*), minimal environmental impact, and low *GWP*, to meet the environmental, health, and safety requirements.

The most important of these characteristics is the dielectric strength, or electric field strength (E_{cr}), which is defined as the maximum voltage that an insulating component can withstand before suffering electrical breakdown [17]. Currently, the different gases that are being researched are hydrofluorocarbons (HFC), perfluorocarbons (PFC), fluoroketones (FK), fluoronitriles (FN), and hydrofluoroolefins (HFO) [15,16]. HFCs and PFCs have excellent dielectric properties [2], but their *GWP* is high, between 5000 and 12,000 in a 100-year horizon [16], making them not suitable environmentally. FKs and FNs have better dielectric properties than SF_6 [15]. One of the most researched FK, $C_5F_{10}O$, has almost the same *GWP* as CO_2, but its boiling point reaches 300 K. Although FNs have higher *GWP*, they present a lower boiling point, high stability, and material compatibility [18]. Finally, a certain researched HFO, $C_3F_4H_2$, has slightly lower dielectric strength than SF_6, a low boiling point, and low *GWP* [16].

Table 2 presents some of the gases and gas mixtures that have been researched for the replacement of SF_6 in medium and high voltage electrical switchgear, and their relative breakdown strength (E_{rel}), *GWP*, boiling temperature (T_B), lifetime (*LT*) in the atmosphere, toxicity threshold limit value–time weighted average (*TLV-TWA*), and decomposition products. The boiling temperature of all these gases is higher than the minimum operating temperature of outdoor medium voltage switchgear of 248 K, according to IEC 62271-200, except C_3F_8, so the gases must be diluted with a carrier gas to ensure that they do not liquefy in any circumstance. The most common carrier gases are the natural gases, CO_2, N_2, and dry air. The dielectric strength of the components is similar to that of SF_6, even after mixing them with the carrier gas, and the highest *GWP* is 70% lower. Besides, the lifetime in the atmosphere is lower than 20 days for the components CF_3I, $C_3F_4H_2$, $C_5F_{10}O$, and $C_6F_{12}O$. Furthermore, the TLV-TWA values of all the gas mixtures are lower than the one of SF_6; however, as they need to be mixed with a carrier gas, the toxicity decreases.

The key to identifying what could be a suitable alternative is to find a component with low *GWP* that preserves a high dielectric strength when it is diluted with a vector gas to reduce the boiling temperature. Various researchers have already presented patents for some of these gas mixtures for their use in medium and high voltage switchgear [19–22].

Table 2. Relative breakdown strength (E_{rel}), global warming potential values in a 100-year horizon (GWP_{100}), boiling point (T_B), toxicity threshold limit values–time weighted average (*TLV-TWA*), and decomposition products of the main researched alternative gas or gas mixtures to replace SF_6 in medium and high voltage switchgear.

Component	Mixture	E_{rel}	GWP_{100}	T_B (K)	LT (yr)	TLV-TWA (ppm_v)	Decomposition Products	Ref.
Sulfur hexafluoride (SF_6)	Pure	1	23,500	209	3200	1000	COF_2, F_2, HF, H_2S, NF_3, F_2O, SO_2, S_2F_{10}, SF_4, SO_2F_2	[1,23,24]
Trifluoroiodomethane (CF_3I)	Pure 30% CF_3I/CO_2	1.2 0.8	<5	251	0.005	150	CF_4, C_2F_6, C_2F_4, C_3F_8, C_3F_6, C_4F_8, C_2F_5I, C_3F_7I, $C_2F_6O_3$, C_3F_7IO	[2,25–27]
Trifluoromethanesulphonyl fluoride (CF_3SO_2F)	Pure 50% CF_3SO_2F/N_2	1.4 1	3678	251	40	*	No products after alternating current (AC) voltage breakdown	[28–30]
Octafluoropropane (C_3F_8)	Pure 20% C_3F_8/N_2	0.9 0.6	7000	236	2600			[31–33]
Heptafluorobutyronitrile (C_4F_7N)	Pure 5% C_4F_7N/N_2 20% C_4F_7N/CO_2	2 0.9 1	2100	268	47	65	CF_4, C_2F_6, C_3F_6, C_3F_8, C_2F_4, $F_3CC{\equiv}CCF_3$, $CF_3CF{=}CFCF_3$, $(CF_3)_3CF$, CF_3CN, C_2F_5CN, CNCN, HF, HCN, CO, CF_3CN, C_2F_6	[34–37]
1,3,3,3-Tetrafluoropropene ($C_3F_4H_2$)	Pure	0.8	6	254	0.05	1000	CF_3CCH, CF_3CCF, C_2H_2, C_2HF, CF_4, HF, CF_3H, C_2F_6, C_2F_4, C_2HF_5, CF_3HCF_3H, C_3F_8	[23,38–40]
Heptafluoroisopropyl trifluoromethyl ketone ($C_5F_{10}O$)	Pure 5% $C_5F_{10}O/Air$	2 0.6	<1	300	0.04	225	CF_4, C_2F_6, C_3F_8, C_3F_6, C_4F_{10}, CF_2O	[15,41–44]
Perfluoro-2-methyl-3-pentanone ($C_6F_{12}O$)	Pure	2.7	<1	322	0.02	150		[15,41,45]

* Safety data sheet for CF_3SO_2F states that the component is fatal if swallowed, inhaled, or in contact with skin.

Binary and ternary mixtures made with the hydrofluoroolefin $C_3F_4H_2$, also known as HFO-1234ze(E), and the perfluoroketone $C_5F_{10}O$, have been studied. They will be referred to as HFO3E and PFK5, respectively, and their molecular structures are shown in Figure 1. These gases have been selected by Ormazabal Corporate Technology. Due to their boiling point being higher than that of SF_6, as shown in Table 2, dry air has been used as a carrier gas. Because of that, the dielectric strength of the gas mixture is also reduced, so a balance must be found between getting the boiling point low enough, while maintaining the dielectric properties. The boiling point of PFK5 is higher than the one of HFO3E, which means that it should be more diluted. Moreover, this favors the fact that HFO3E has lower dielectric strength.

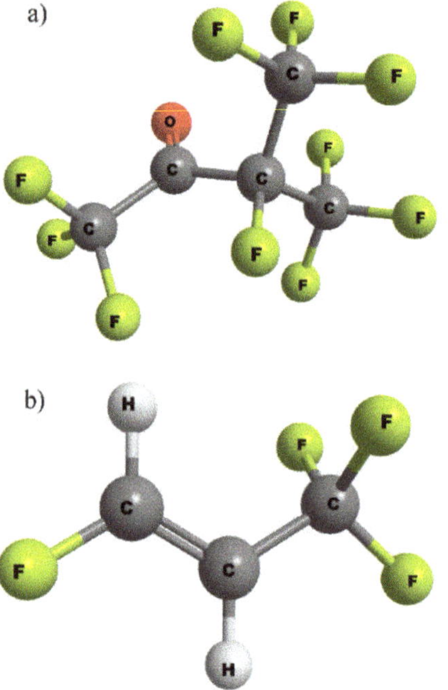

Figure 1. Molecular structures of (**a**) $C_5F_{10}O$ (PFK5) and (**b**) $C_3F_4H_2$ (HFO3E).

The understanding of the mixing process of these new gas mixtures is critical to define the physical and chemical properties that affect the electrical insulation and arc quenching properties. Therefore, the goal of this paper is the study of the diffusional behavior of the binary and ternary mixtures made with PFK5, HFO3E, and dry air, as possible alternatives for SF_6 for medium voltage switchgear, to anticipate the oncoming change of regulations. A chamber has been designed and built to study the diffusion of these binary and ternary mixtures, using UV-Vis spectroscopy to measure the concentration in the chamber during the mixing process. This technique was already used by some authors to measure these kinds of gases [46,47].

A diffusion process model is proposed to describe the mixture process; the model is fitted to the experimental concentration values in an algorithm developed in Scilab, which numerically calculates the diffusion parameters of the mixtures.

In this work, a novel experimental equipment capable of measuring and collecting the online concentration of a gas mixture has been designed and tested with fluorinated gas mixtures, candidates to replace SF_6 in medium voltage switchgear. Although this study has been focused on the analysis of a specific case of new fluorinated gas mixtures, this non-destructive technology would be of great interest for any application in which it is necessary to determine the physicochemical properties of a gas mixture through its composition. Some examples include the design of gas mixtures that require specific characteristics, the monitoring of reactive mixtures, or the online malfunction detection and maintenance without the need to take samples that could compromise the integrity of the gas mixtures.

Moreover, a mathematical model that simulates the diffusion process of a component in a binary or ternary gas mixture has been proposed and implemented. The mixing process of the new insulating gas mixtures and the time it takes for them to reach stability have been determined.

This article intends to lay the groundwork for future experiments to study the effect of different perturbations, such as temperature, humidity, and concentration changes, among others, during the filling, mixing, and post-mixing process of the new insulating gas mixtures.

2. Materials and Methods

2.1. Experimental Equipment

The gas diffusion was determined using an experimental system that was specifically designed for this project to contain the gas mixtures and measure their concentration over time, using UV-Vis spectroscopy technology. The experimental equipment is shown in Figure 2.

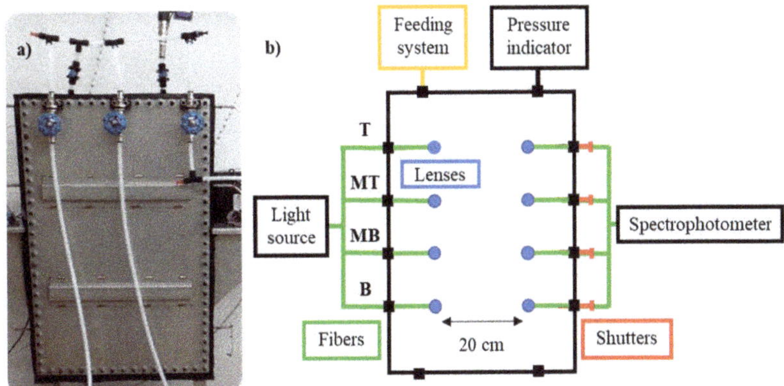

Figure 2. (**a**) Gas chamber and (**b**) scheme of the experimental equipment used for the gas diffusion study and its components. T (Top), MT (Medium Top), MB (Medium Bottom), and B (Bottom) correspond to the different heights of the measurement lines.

The designed system consisted of a 60 L tank, a DH-2000-S-DUV-TTL light source (Ocean Insight, Orlando, FL, USA), a Maya2000Pro spectrophotometer (Ocean Insight), a pressure indicator (WIKA, Klingenberg, Bayern, Deutschland), and a feeding system composed of polytetrafluoroethylene (PTFE) tubes to introduce the gas mixtures in the tank. The tank had similar dimensions to an actual medium voltage switchgear, but was slightly narrower, $0.8 \times 0.5 \times 0.15$ m, which favors vertical gas diffusion.

The light source emits a continuous UV-Vis spectrum, between 190–750 nm, which is split into the four fibers that end in four lenses that are located at four different heights of the tank, as the interest is focused on the diffusion that happens vertically. The measurement lines are located at 0.13, 0.35, 0.57, and 0.78 m; for simplicity reasons, the lines are referred to as Bottom (B), Medium Bottom (MB), Medium Top (MT), and Top (T), respectively.

The emitted light passes through the tank and is collected in four other lenses and fibers that merge and terminate in the spectrophotometer, which provides the corresponding ultraviolet absorbance spectra. A shutter is installed in each line, so the signal of each line is measured individually.

2.2. Mixing Process Tracking

The concentration (C) of the components in the gas mixtures that were studied are shown in Table 3. Each gas mixture was replicated three times. The components of the mixture were introduced in the tank one at a time through the feeding system, according to their molecular weight. An initial state of stratification by density was created by inserting the gases from heaviest to lightest, favoring the monitoring of the mixing process.

Table 3. Nomenclature and concentration (C) of the components of the studied gas mixtures.

Name	C_{PFK5} (%)	C_{HFO3E} (%)	$C_{DRY\ AIR}$ (%)
B-10P	10	0	90
B-20P	20	0	80
B-40H	0	40	60
T-10P/40H	10	40	50

The dielectric strength, or critical field strength (E_{cr}), of the gas mixtures mentioned above can be estimated by a linear scaling according to Saxegaard et al. [48]:

$$E_{cr} = \sum_{i}^{n} x_i E_{cr,i} \left[\text{V m}^{-1}\,\text{Pa}^{-1} \right] \tag{1}$$

where x_i and $E_{cr,i}$ are the molar fraction and the electrical field strength of each component of the mixture, taking into consideration the dielectric strength of SF_6 and dry air (89 and 27 V m^{-1} Pa^{-1}) [49], and the E_{rel} values of PFK5 and HFO3E that are shown on Table 2. The E_{cr} of components PFK5 and HFO3E are 178 and 71 V m^{-1} Pa^{-1}, and the estimated values are 42, 57, 37, and 53 V m^{-1} Pa^{-1} for the gas mixtures B-10P, B-20P, B-40H, and T-10P/40H, respectively.

The measurement process of absorbance was automatized using LabVIEW, identifying the values of the absorbance for the 196 and 300 nm wavelengths, which correspond to the maximum absorption of HFO3E and PFK5, respectively. The system was programmed to open and close the shutters of the light source and the fibers, and to control the spectrophotometer to measure the reference intensities and calculate the absorbance of the four lines sequentially. The absorbance was calculated every 15 min for the first hour, every 20 min for the second hour, and every 30 min until the end of the experiment, which ends after 20 h.

Absorbance (A) was calculated by the Beer–Lambert Law [50]:

$$A = \log\left(\frac{I_0 - D_0}{I - D_0} \right) = \varepsilon_M\, C\, l\,[\text{AU}] \tag{2}$$

where I_0 is the reference intensity, I is the intensity that is read after the absorption, and D_0 is the dark intensity, the one that the spectrophotometer reads when no light is emitted. A can also be expressed by the molar absorptivity (ε_M, m^3 mol^{-1} m^{-1}) multiplied by the molar concentration of the species (C) and the path length (l).

The mean value of the absorbance that is obtained in the three replicas of each experiment was the one used for the numerical calculation of the diffusion parameters.

2.2.1. Determination of the Concentration

The concentration was obtained through a calibration that relates it with the absorbance of each component. Four gas mixtures of known concentration were prepared for each component: 5%, 10%, 15% and 20% mixtures for PFK5, and 10%, 20%, 30% and 40% for HFO3E.

Each mixture was introduced in the gas chamber and the absorbance was monitored until the mixing process was completed, which means when all measurement lines displayed a similar value. The absorbance value that was obtained is the one related to the specific concentration that was introduced.

The *A-C* data were plotted and adjusted to linear regression to calculate the equations of the calibration curves. These equations allowed the calculation of the concentration of the components from the measurements of absorbance.

2.2.2. Diffusional Model

Two effects needed to be studied to determine the variation of concentration over time: (a) molecular diffusion, the transfer of individual molecules due to a concentration, pressure, or temperature gradient, among others; and (b) convection transport, caused by the overall movement of the fluid [51]. As the gases used in these experiments have high molecular weight, it was considered that the gas mixing process happened not only due to the concentration gradient, but also to the differences in the density of the components or natural convection.

Considering the described two effects, a modified Fick's generalized expression was obtained:

$$\frac{dC}{dt} = D\,\frac{\partial^2 C_i}{\partial z^2} + \frac{\partial(V_i\,C_i)}{\partial z} \tag{3}$$

where C_i is the concentration, D (m^2 s^{-1}) is the diffusion coefficient, t represents the time, and z the height. V_i is the velocity caused by the differences between densities:

$$V_i = K\,(\rho_m - \rho_i)\left[\text{m s}^{-1}\right] \tag{4}$$

where K (m^4 kg^{-1} s^{-1}) is the convection constant. The density of the heavy component in the binary mixture (ρ_m) and the density of the mixture at each point in time (ρ_i) were calculated by:

$$\rho_m = \frac{P\,MW_m}{R\,T}\left[\text{kg m}^{-3}\right] \tag{5}$$

$$\rho_i = \frac{P\,MW_i}{R\,T}\left[\text{kg m}^{-3}\right] \tag{6}$$

T and P are the temperature and pressure at which the experiments were carried out, 298 K and 1 atm (101,325 Pa), and R is the ideal gas constant. MW_m is the molecular weight of the heavy component of the binary mixture, and MW_i indicates the molecular weight of the mixture at each point in time, which was calculated by the weighted average of the MW_m of each component of the mixture:

$$MW_i = \sum_{c}^{n} y_m\,MW_m\left[\text{kg mol}^{-1}\right] \tag{7}$$

The equations were based on the fact that for a binary mixture consisting of components A and B, the diffusivity of A in B (D_{AB}), and vice versa (D_{BA}), are the same [51]. Both molecular diffusion and natural convection were considered to determine the molar flux (J):

$$J = -D\,\frac{dC_i}{dz} + V_i\,C_i\left[\text{mol m}^{-2}\text{ s}^{-1}\right] \tag{8}$$

Theoretical studies of multicomponent diffusion are based on the Stefan–Maxwell equation, which, in turn, was derived from the solution of the Boltzmann equation [52]. For isothermal and isobaric conditions, this equation is a strong approximation that satisfies the practical requirement. The variation of concentration over time for multicomponent mixtures is represented in the following equation:

$$\left(\frac{dC}{dt}\right) = [D]\left(\frac{\partial^2 C_i}{\partial z^2}\right) + \left(\frac{\partial(V_i\,C_i)}{\partial z}\right) \tag{9}$$

The element $[D]$ is a size $n - 1$ squared matrix that represents the diffusion coefficients, n being the number of components in the mixture. The elements of the square matrix $[D]$

are called practical diffusion coefficients [52]. In this study, a gas mixture composed of three elements was made; therefore, the matrix was composed of 2×2 elements. As well as in binary gas mixtures, the convection effect was also considered in multicomponent mixing:

$$\begin{pmatrix} \frac{dC_1}{dt} \\ \frac{dC_2}{dt} \end{pmatrix} = \begin{bmatrix} D_{11} & D_{12} \\ D_{21} & D_{22} \end{bmatrix} \begin{pmatrix} \frac{\partial^2 C_1}{\partial z^2} \\ \frac{\partial^2 C_2}{\partial z^2} \end{pmatrix} + \begin{pmatrix} \frac{\partial(V_1\,C_1)}{\partial z} \\ \frac{\partial(V_2\,C_2)}{\partial z} \end{pmatrix} \tag{10}$$

Hence, the concentration differential of each component over time depends on the practical diffusion coefficients of both components and their concentration differential over the position in the gas chamber [52]:

$$\frac{dC_1}{dt} = D_{11}\frac{\partial^2 C_1}{\partial z^2} + D_{12}\frac{\partial^2 C_2}{\partial z^2} + \frac{\partial(V_1\,C_1)}{\partial z} \tag{11}$$

$$\frac{dC_2}{dt} = D_{21}\frac{\partial^2 C_1}{\partial z^2} + D_{22}\frac{\partial^2 C_2}{\partial z^2} + \frac{\partial(V_2\,C_2)}{\partial z} \tag{12}$$

Furthermore, the effective diffusivity coefficients (D_{eff}) were calculated using the practical diffusion coefficients [52]. Bird et al. [53] stated that the molar fraction gradients could be replaced by molar fraction differences to simplify the calculations:

$$D_{1,eff} = \sum_{k=1}^{n-1} D_{1k}\frac{\nabla y_k}{\nabla y_1}\left[\text{m}^2\,\text{s}^{-1}\right] \tag{13}$$

Furthermore, according to Taylor and Krisna [52], the diffusivity eigenvalues of the components ($\hat{D}_1$ and $\hat{D}_2$) must be positive:

$$\hat{D}_1 = \frac{1}{2}\left(tr[D] + \sqrt{disc[D]}\right) \tag{14}$$

$$\hat{D}_2 = \frac{1}{2}\left(tr[D] - \sqrt{disc[D]}\right) \tag{15}$$

where $tr[D]$ is the trace and $disc[D]$ is the discriminant of the matrix. Because of Equations (14) and (15), the conditions presented in the following equations must be met:

$$D_{11} + D_{22} > 0 \tag{16}$$

$$D_{11}\,D_{22} - D_{12}\,D_{21} > 0 \tag{17}$$

$$(D_{11} - D_{22})^2 + 4\,D_{12}\,D_{21} > 0 \tag{18}$$

The proposed mathematical model was reproduced in an algorithm created in Scilab. The purpose of the program is the numerical calculation of the differential Equations (3), (11) and (12) to obtain the optimized diffusion and convection coefficients by minimizing an error objective function (OF):

$$OF = \frac{\sum_1^{4N}\left(C_{exp} - C_{model}\right)^2}{4N} \tag{19}$$

Such OF was defined as the sum of the squared differences between the experimental values of the concentration (C_{exp}) and the ones obtained in the model (C_{model}) divided by the whole amount of measured concentrations in the lines (N). The relative standard error (RSE) [54] between the experimental concentration values and the model was also calculated by referencing them to the final concentration value (C_{final}):

$$RSE = \sqrt{\frac{\sum_1^{4N}\left(\frac{C_{exp} - C_{model}}{C_{final}}\right)^2}{4N - 1}}\ 100\ [\%] \tag{20}$$

The optimization algorithm looked for the values of D and K in the differential equations so that the theoretical model fitted the experimental values and also calculated the relative standard error.

Additionally, the binary diffusion coefficients (D_E) were also calculated using the correlation proposed by Fuller, Schettler, and Giddings [52,55]:

$$D_E = D_{AB} = 0.01883 \, T^{1.75} \frac{\sqrt{\frac{(M_A + M_B)}{M_A \, M_B}}}{P \left[\sqrt[3]{\sum v_A} + \sqrt[3]{\sum v_B} \right]^2} \left[\text{m}^2 \, \text{s}^{-1} \right] \qquad (21)$$

The results were compared with the numerically obtained optimized diffusion constants of all binary mixtures. The binary diffusion coefficient between components A and B (D_{AB}) depends on the temperature (T, K), pressure (P, Pa), molecular weight (M, g mol^{-1}), and atomic diffusion volumes (v). The values of v for the atoms C, H, O, and F are 15.9, 2.31, 6.11, and 14.7, respectively [56].

The values of D_E of the ternary mixture were calculated by dividing the mixture into two pseudo-binary mixtures between one heavy component against the other two components as a whole. The estimated value for each heavy component was compared with the numerically calculated D_{eff} values.

3. Results and Discussion

3.1. Mixing Process Tracking

3.1.1. Determination of the Concentration

The calibration curve of each component and measurement line is shown in Figure 3. The curves present a linear tendency. The linear regression equations, correlation coefficients (r^2), and the molar absorptivity of PFK5 and HFO3E were calculated, and they are collected in Table 4. The results show that the sensitivity of the measuring system for HFO3E is considerably lower than the one of PFK5, due to the great difference in the molar absorptivity of both gases.

The maximum absorption of the component HFO3E is located at 196 nm in the UV-Vis spectrum, where it is common to find noise when measuring the absorbance since this wavelength shows very poor selectivity. Because of this, the concentration, velocity and molar flux profiles of HFO3E have more deviations and are less accurate.

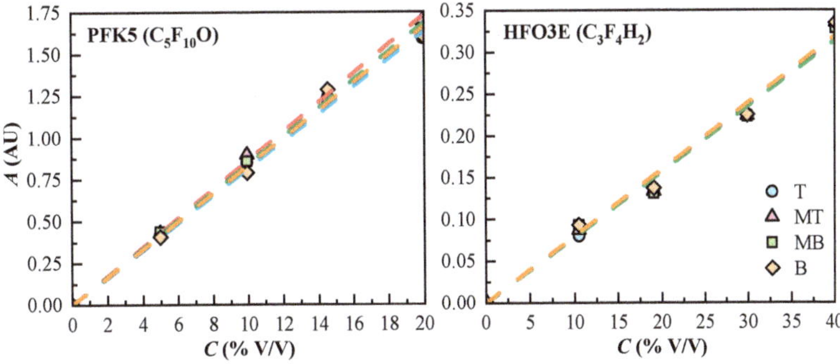

Figure 3. Experimental concentration values (C_{exp}, dots) and calibration curves (dashed lines) of the components of the gas mixtures. T (Top, $\bigcirc$), MT (Medium Top, $\triangle$), MB (Medium Bottom, $\square$), and B (Bottom, $\diamondsuit$) correspond to the different heights of the measurement lines.

Table 4. Linear regression equations, correlation coefficients (r^2), and molar absorptivities (ε_M), of PFK5 and HFO3E.

	Line	Linear Regression	r^2	ε_M (m^3 mol^{-1} cm^{-1})
	T	$y = 0.0811x$	0.997	9.9
PFK5	MT	$y = 0.0852x$	0.999	10.4
	MB	$y = 0.0827x$	0.999	10.1
	B	$y = 0.0826x$	0.994	10.1
	T	$y = 0.0081x$	0.991	0.99
HFO3E	MT	$y = 0.0081x$	0.989	0.99
	MB	$y = 0.0079x$	0.988	0.97
	B	$y = 0.0081x$	0.989	0.98

3.1.2. Diffusional Model

The concentration profiles of the model and the diffusional coefficients were obtained thanks to the optimization process in Scilab. The concentration profiles of the binary mixtures are plotted in Figure 4. The 95% confidence interval of the experimental values is indicated. The B-40H mixture showed broader confidence intervals than the binary mixtures of PFK5. The concentration of all binary mixtures reached stabilization after 10 h. The molar concentrations at the end of the experiments were 4.1 and 8.2 mol m^{-3} for the PFK5 mixtures and 17.1 mol m^{-3} for the HFO3E mixture.

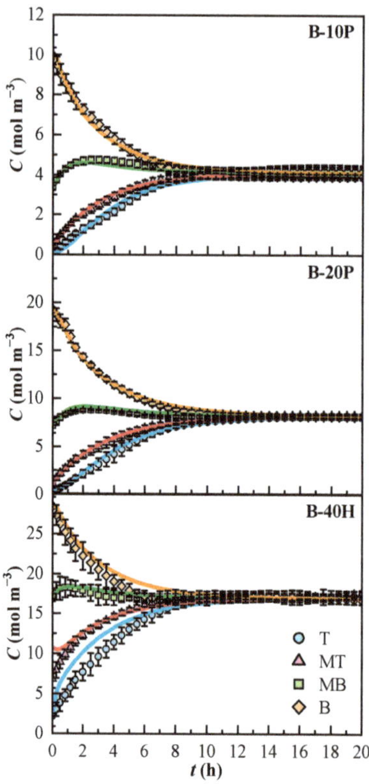

Figure 4. Experimental (C_{exp}, dots) and estimated (C_{model}, solid lines) concentration profiles of the binary mixtures over time. T (Top, ○), MT (Medium Top, △), MB (Medium Bottom, □), and B (Bottom, ◇) correspond to the different heights of the measurement lines.

The model fitted the experimental data of B-10P and B-20P mixtures correctly. The concentration profiles from the mixture B-40H show that the model deviated in lines T and B at the beginning of the experiment. The molar absorptivity of HFO3E is ten times lower than that of PFK5 in the UV spectrum. T is the line that shows the lowest concentration and B is the line that shows the highest molar flux, as will be discussed later.

The concentration profiles obtained for the T-10P/40H mixture are shown in Figure 5. In general, the ternary mixture showed greater confidence intervals than the binary mixtures. Line MB exhibited less variation in both components than the others. Even though the whole mixture stabilized after 12 h, the concentration of HFO3E was stable after 10 h. The compound PFK5 is the one that determines the duration of the mixing process, as it is the one with the lower effective diffusion coefficient. The molar concentrations at the end of the experiments were 4.4 mol m^{-3} for PFK5 and 16.6 mol m^{-3} for HFO3E. The model fitted the experimental data correctly, although the concentration profile of HFO3E also deviated from the confidence interval in lines T and B, like in mixture B-40H.

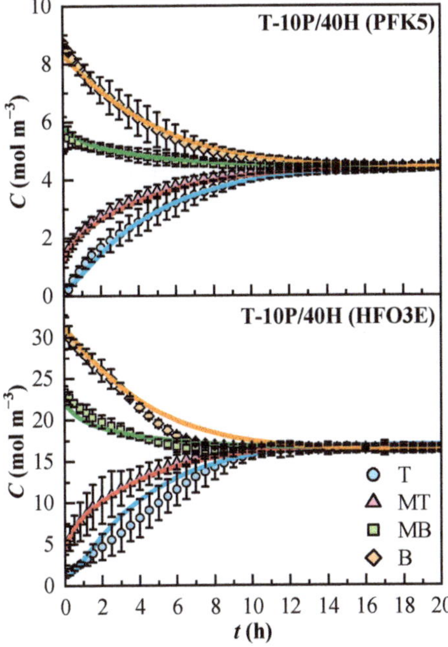

Figure 5. Experimental (C_{exp}, dots) and estimated (C_{model}, solid lines) concentration profiles of the components of the ternary mixture over time. T (Top, ○), MT (Medium Top, △), MB (Medium Bottom, □), and B (Bottom, ◇) correspond to the different heights of the measurement lines.

The optimized diffusion and convective coefficients, the theoretically estimated coefficients, and relative standard error for the binary mixtures are shown in Table 5. As diffusion does not depend on concentration, the D of mixtures B-10P and B-20P should be equal, and they are similar. Additionally, the K values are very low. It can be assumed that the convection mechanism had little effect on the mixing process. The RSE for all binary mixtures is lower than 6%, and the mixture whose model best fitted the experimental data was B-20P, which also has the lowest diffusion coefficient. The optimized coefficients of the PFK5 mixtures are more similar to the estimated one than the HFO3E mixture, and the error increases with the concentration.

Table 5. Diffusion (D_E and D) and convective (K) coefficients and relative standard error (RSE) of the binary mixtures after the optimization process using Scilab.

Mixture	D_E * (m^2 s^{-1})·10^6	D (m^2 s^{-1})·10^6	K (m^4 kg^{-1} s^{-1})·10^{14}	RSE (%)
B-10P	5.4	5.2	3.1	3.8
B-20P	5.4	4.8	2.7	2.6
B-40H	7.9	5.3	2.7	5.3

* Theoretically estimated using Equation (21).

For the ternary mixture, the values of the practical diffusion coefficients were the following:

$$[D] = \begin{pmatrix} D_{11} & D_{12} \\ D_{21} & D_{22} \end{pmatrix} = \begin{pmatrix} 0.0195 & -0.0017 \\ 0.0535 & 0.0001 \end{pmatrix} \tag{22}$$

The conditions that Taylor and Krisna described [52], which are shown in Equations (16)–(18), were met.

The diffusion and convection coefficients and RSE values are shown in Table 6. Both values of K are also very low in the ternary mixture. The D_{eff} for PFK5 and HFO3E are calculated from Equation (13). They are lower than the D obtained for the binary mixtures, which states that the ternary mixture diffuses slower than the binary mixtures. The D_{eff} for HFO3E is higher than the D_{eff} for PFK5. That of HFO3E is closer to its estimated value than the one of PFK5. HFO3E diffuses slower and PFK5 diffuses faster than what was theoretically estimated. Overall, the model fitted better the experimental data of PFK5 than the data of HFO3E. The results are consistent with what is shown in Figure 5.

Table 6. Diffusion ($D_{eff,E}$ and D_{eff}) and convective (K) coefficients and relative standard error (RSE) of the components of the ternary mixture after the optimization process in Scilab.

Component	$D_{eff,E}$ * (m^2 s^{-1})·10^6	D_{eff} (m^2 s^{-1})·10^6	K (m^4 kg^{-1} s^{-1})·10^{11}	RSE (%)
PFK5	2.9	3.6	0.6	2.2
HFO3E	4.4	3.9	2.2	4.4

* Theoretically estimated using Equation (21).

The convection velocity of the binary mixtures is shown in Figure 6. The velocity is negative, as K is positive and the mixture with air has lower density than PFK5 or HFO3E. The convection phenomenon pushes down the heavy components of the mixtures. The most negative velocity values are found in line T, which is the highest line in the gas chamber and therefore the most affected by the high density of the components. The velocities are less negative as the height in which the measurement line is located decreases. Nonetheless, the convection velocity is so small that it is almost negligible.

The convection velocity of T-10P/40H is shown in Figure 7. As the component PFK5 is the heaviest, its convection velocity is always negative, because any mixture between HFO3E and air will always have lower density than PFK5. The convection velocity of component HFO3E is positive in line B at the beginning of the experiment, which happens because the higher concentration of PFK5 pushes up HFO3E and the carrier gas.

The molar flux of the binary mixtures is shown in Figure 8. All binary mixtures have a similar tendency. The molar flow is higher in line B, the one that presents the highest concentration values. The molar flow is almost zero in line T. The molar flow of lines MT and MB slightly increase at the beginning of the experiments.

The molar flux of the ternary mixture can be observed in Figure 9. Both molar flux profiles share the same tendency, are very similar, and reach zero approximately at the same time. The lines that present higher molar flux are MT and MB, as they show the highest concentration gradients between all components. The molar flux of line B is the one that increased at the beginning of the mixing process, and line T is also almost zero.

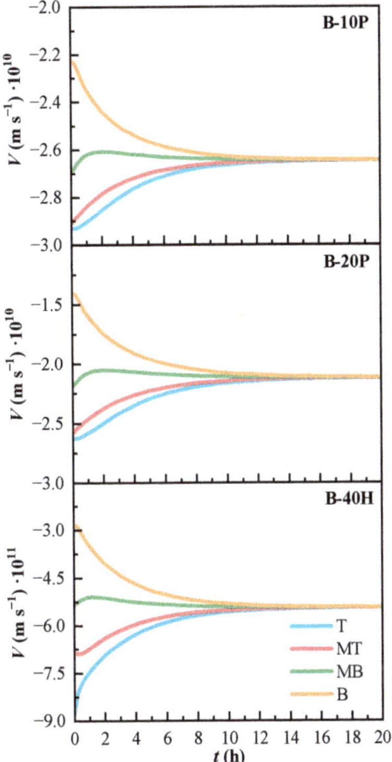

Figure 6. Variation of the estimated convection velocity (V) of the binary mixtures over time. T (Top), MT (Medium Top), MB (Medium Bottom), and B (Bottom) correspond to the different heights of the measurement lines.

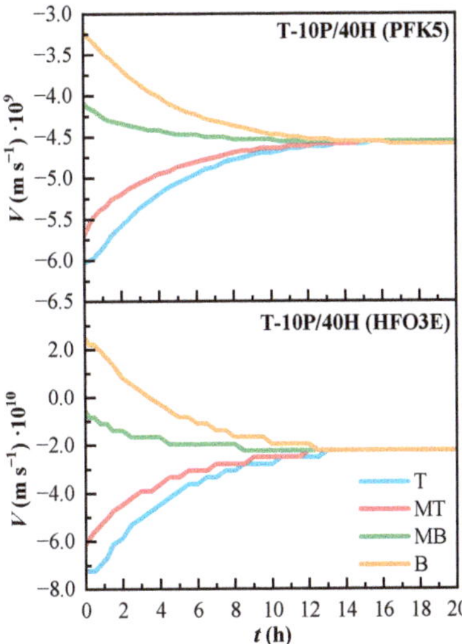

Figure 7. Variation of the estimated convection velocity (V) of the components of the ternary mixture over time. T (Top), MT (Medium Top), MB (Medium Bottom), and B (Bottom) correspond to the different heights of the measurement lines.

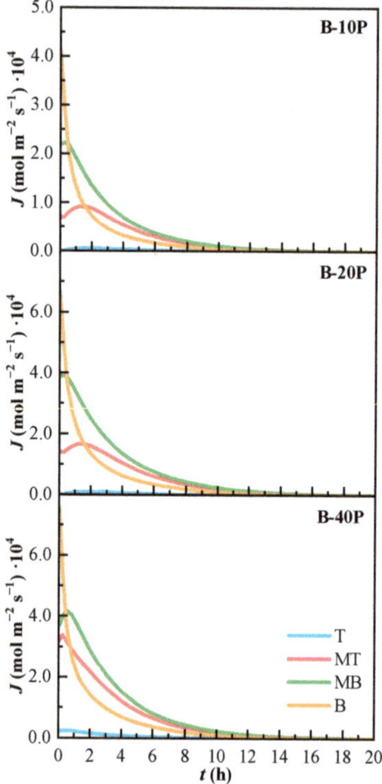

Figure 8. Variation of the estimated molar flux (*J*) of the binary mixtures over time. T (Top), MT (Medium Top), MB (Medium Bottom), and B (Bottom) correspond to the different heights of the measurement lines.

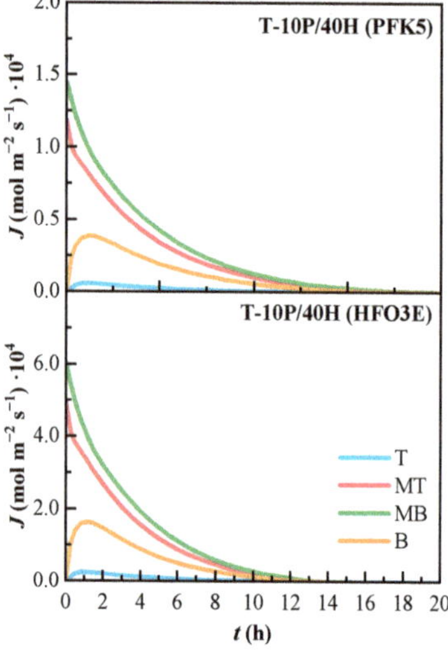

Figure 9. Variation of the estimated molar flux (*J*) of the components of the ternary mixture over time. T (Top), MT (Medium Top), MB (Medium Bottom), and B (Bottom) correspond to the different heights of the measurement lines.

The molar flux profiles, together with the obtained concentration and velocity profiles, represent that all the mixtures reached homogeneity and stability at the end of the experiments.

3.1.3. Model Validation

The validation of the model was carried out to determine its accuracy not only for the gas mixtures used to build it, but also for mixtures made with other concentration values too [53]. Five gas mixtures were proposed, different from the ones used to prepare the model. The composition of the gas mixtures is shown in Table 7.

Table 7. Nomenclature and concentration (C) of the components of the gas mixtures used for validation.

Name	C_{PFK5} (%)	C_{HFO3E} (%)	$C_{DRY\,AIR}$ (%)
B-5P	5	0	95
B-15P	15	0	85
B-15H	0	15	85
B-30H	0	30	70
T-20P/40H	20	40	50

The concentration profiles of these mixtures were simulated. The D and K coefficients that were used for the binary mixtures are the ones shown in Table 5, and the D_{eff} and K coefficients that were used for the ternary mixture are obtained from Table 6. The residuals of each mixture, throughout the whole experiment, were calculated:

$$Residual = C_{exp} - C_{model} \left[\text{mol m}^{-3} \right] \tag{23}$$

The mean residual value of all four measurement lines at a specific time ($\overline{R_t}$) and the mean residual value of a specific measurement line ($\overline{R_L}$) until the end of the experiment (t_f) were also calculated:

$$\overline{R_t} = \frac{\sum_{L=1}^{4}\left(C_{exp} - C_{model}\right)_L}{4} \left[\text{mol m}^{-3} \right] \tag{24}$$

$$\overline{R_L} = \frac{\sum_{t=0}^{t_f}\left(C_{exp} - C_{model}\right)_t}{N} \left[\text{mol m}^{-3} \right] \tag{25}$$

The end of the experiment was considered when the concentration of the four lines reached stabilization. All the mean residual values were represented against C_{exp}, including the mixtures that were used for building the model [53]. $\overline{R_t}$ is represented against C_{exp} in Figure 10. The residuals of PFK5 do not show a distinct tendency; they are distributed almost evenly in the positive and negative sides of the y-axis and are near zero. This shows that the error of PFK5 is not carried over with the model. There is no significant difference between the $\overline{R_t}$ of PFK5 of binary and ternary mixtures.

The results for HFO3E show higher residual values and they are not evenly distributed. Most of them are negative values, which shows that the model is more likely to present higher values than the experimental ones. The residuals also decrease with time, so the model showed better results near stabilization. The low absorptivity of this component, together with the higher molar flow at the beginning of the experiments, makes it more difficult to predict the values of concentration. Besides, the concentration of HFO3E of the binary mixtures is easier to calculate as the residuals are lower.

$\overline{R_L}$ is represented against C_{exp} in Figure 11. Lines T and MB show lower residual values for both components. PFK5 clearly shows lower residual values in both binary and ternary mixtures. The values of HFO3E of the ternary mixtures are higher than the ones of the binary mixtures. The model will predict lower values than the experimental ones in line MT and higher ones in line B. Line B shows the highest absolute residual values, especially for HFO3E. The higher molar concentration and molar flux that was found in this line complicated the ability of the model to predict the concentration values.

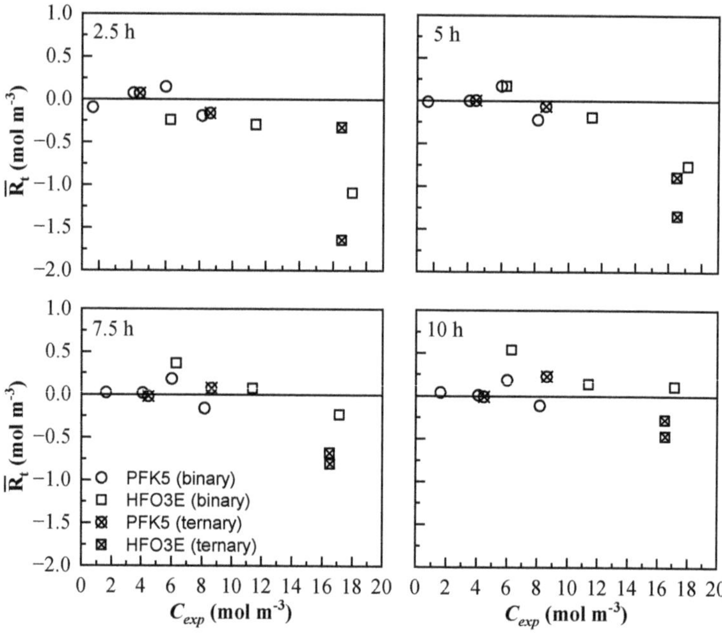

Figure 10. Mean residual value of all measurement lines at different times ($\overline{R_t}$) of the mixing process. The results of all the binary ($\bigcirc$ and $\square$) and ternary ($\otimes$ and $\boxtimes$) mixtures are represented for PFK5 and HFO3E, respectively.

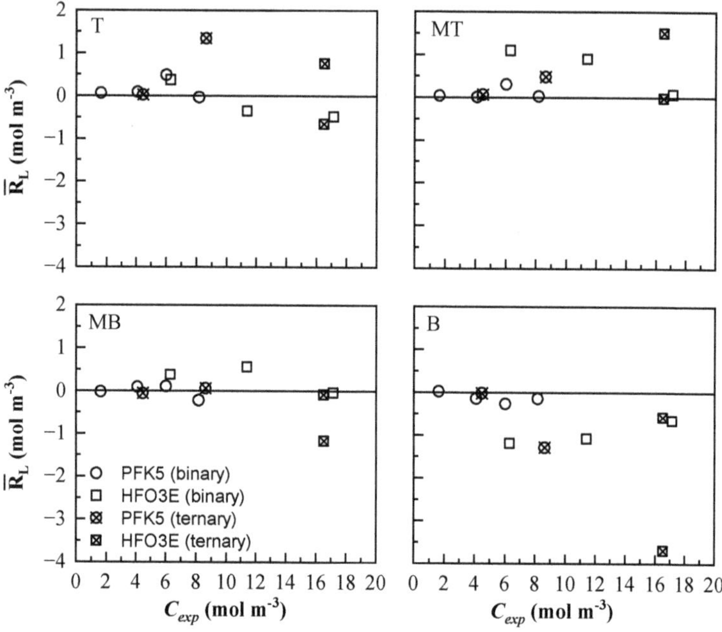

Figure 11. Mean residual value of the measurement lines at all times ($\overline{R_L}$) of the mixing process. The results of all the binary ($\bigcirc$ and $\square$) and ternary ($\otimes$ and $\boxtimes$) mixtures are represented for PFK5 and HFO3E, respectively.

The minimum and maximum absolute residual values for all mixtures are shown in Table 8. As all gas mixtures reach stability after 10 h, only results before that were considered. The maximum residual values of PFK5 were found in line B at the beginning of the experiments. The minimum ones were distributed into all lines.

Table 8. Minimum and maximum absolute residual values of the mixtures.

Mixture		Minimum Absolute Residuals			Maximum Absolute Residuals		
		Value·10^3 (mol m^{-3})	t (h)	Line	Value (mol m^{-3})	t (h)	Line
PFK5	B-10P	9	0.5	MT	0.34	0.5	B
	B-20P	0.83	1.5	T	0.73	1	B
	T-10P/40H	2	8	T	0.34	0.5	B
	B-5P	0.33	5.5	B	0.28	2.5	T
	B-15P	3	2.5	MB	1.88	0.5	B
	T-20P/40H	3.4	8.5	MB	4.09	0.5	B
HFO3E	B-40H	2.4	8	MB	2.73	1	T
	T-10P/40H	20	10	T	2.25	3.5	T
	B-15H	40	7.5	T	3.57	2.5	B
	B-30H	300	10	MB	3.51	3	B
	T-20P/40H	450	10	T	7.39	3.5	B

HFO3E showed its maximum residuals distributed between lines T and B at the beginning of the experiments. The minimum residual values of HFO3E increase when the concentration increases. The T-20P/40H mixture was the one that showed the highest maximum absolute residuals, which were presented in line B.

4. Conclusions

A novel piece of experimental equipment has been designed, and an automatized measurement system, which has been programmed in LabVIEW, has been implemented for the online monitoring of the mixing process of the gas mixtures.

Consequently, it has been possible to determine the duration of the mixing process of new insulating gas mixtures made of PFK5 and HFO3E, candidates to replace SF_6 in MVS. Starting from an initial stratification state caused by the feeding of the gases depending on their molecular weight, stability in the mixtures has been reached between 10 and 12 h, depending on the composition of the mixture. Besides, binary mixtures have reached stability before the ternary mixture. In addition, it has been seen that the molar absorptivity of HFO3E is lower than of PFK5, and that it absorbs at 196 nm in the UV-Vis spectrum, which has hindered measurements.

A mathematical model that describes the mixing process has been proposed, which takes both natural convection and molecular diffusion into consideration. The model has made it possible to obtain the diffusion and convection coefficients of the mixtures with a relative standard error lower than 6%. In comparison with the calculation of the diffusion coefficient of HFO3E that Hu et al. [57] performed, the estimation shares the same order of magnitude, although it is slightly higher in this work. The convection coefficients of all the mixtures have been significantly low, which could prove that the effect of natural convection is negligible in comparison with molecular diffusion of the order of 10^5 (in the ternary mixture) to 10^8 (in the binary mixtures) times smaller.

The residuals of PFK5, which have shown no significant tendency, neither in sign (positive or negative) nor in value, are lower than 1.9 mol m^{-3} for the binary mixtures and 4.1 for the ternary mixtures. Those of HFO3E have been higher at the beginning of the experiment, between 2.3 and 7.4 mol m^{-3}, and have decreased with time, reaching values as low as 2.4×10^{-3} mol m^{-3}. The model has predicted the concentration values of line MB, for example, better than that of the bottom line, probably due to high variability in the concentration.

It has been possible to predict the concentration profiles of a mixing process of gas mixtures made of PFK5 to be better than the ones made of HFO3E, using air as the carrier gas. The low absorptivity of HFO3E in the UV spectrum has led to higher residual values in the lines that have the highest and lowest concentration of both binary and ternary mixtures. This indicates that UV-Vis could not be the best technique to monitor the concentration changes of this component.

Through this work, the physical characterization of the diffusion process and mixing stability of binary and ternary gas mixtures made of PFK5 and HFO3E was started. All this information is postulated as a methodological and diffusional modeling basis for future studies that take into account other disturbances, such as temperature, humidity, and electrical discharge, among others. Nevertheless, attention also needs to be paid to the chemical stability of the mixtures using analytical methods in future studies.

Author Contributions: Conceptualization, J.I.L.; methodology, A.E. and L.C.; software, E.A., A.E. and B.P.-A.; validation, A.E., M.L.A. and R.M.A.; formal analysis, A.E.; investigation, A.E. and L.C.; resources, J.I. (Jesús Izcara) and J.I. (Josu Izagirre); data curation, A.E.; writing—original draft preparation, A.E. and L.C.; writing—review and editing, J.I.L., M.L.A. and R.M.A.; visualization, J.I.L.; supervision, J.I.L.; project administration, J.I.L., R.M.A. and J.I. (Josu Izagirre); funding acquisition, J.I.L., R.M.A. and J.I. (Josu Izagirre). All authors have read and agreed to the published version of the manuscript.

Funding: This research was funded by the Basque Government, grant numbers KK-2017/00090 and KK-2019/00017.

Institutional Review Board Statement: Not applicable.

Informed Consent Statement: Not applicable.

Data Availability Statement: Not applicable.

Conflicts of Interest: The authors declare no conflict of interest.

References

1. Tsai, W.-T. The decomposition products of sulfur hexafluoride (SF_6): Reviews of environmental and health risk analysis. *J. Fluor. Chem.* **2007**, *128*, 1345–1352. [CrossRef]
2. Xiao, S.; Zhang, X.; Tang, J.; Liu, S. A review on SF_6 substitute gases and research status on CF_3I gases. *Energy Rep.* **2018**, *4*, 486–496. [CrossRef]
3. Stocker, T.F.; Qin, D.; Plattner, G.-K.; Tignor, M.M.B.; Allen, S.K.; Boschung, J. Climate Change 2013: The Physical Science Basis. Intergovernmental Panel on Climate Change. 2013. Available online: https://www.ipcc.ch/report/ar5/wg1/ (accessed on 13 December 2021).
4. United Nations. Kyoto Protocol to the United Nations Framework Convention on Climate Change. 1998. Available online: https://unfccc.int/kyoto_protocol (accessed on 14 December 2021).
5. Hyrenbach, M.; Paul, T.A.; Owens, J. Environmental and safety aspects of AirPlus insulated GIS. In Proceedings of the 24th International Conference & Exhibition on Electricity Distribution, Glasgow, UK, 12–15 June 2017.
6. Weiss, R.F.; Mühle, J.; Salameh, P.K.; Harth, C.M. Nitrogen trifluoride in the global atmosphere. *Geophys. Res. Lett.* **2008**, *35*, L20821. [CrossRef]
7. Morais, A.R.C.; Harders, A.N.; Baca, K.R.; Olsen, G.M.; Befort, B.J.; Dowling, A.W.; Maginn, E.J.; Shiflett, M.B. Phase Equilibria, Diffusivities, and Equation of State Modeling of HFC-32 and HFC-125 in Imidazolium-Based Ionic Liquids for the Separation of R-410A. *Ind. Eng. Chem. Res.* **2020**, *59*, 18222–18235. [CrossRef]
8. Mühle, J.; Ganesan, A.L.; Miller, B.R.; Salameh, P.K.; Harth, C.M.; Greally, B.R.; Rigby, M.; Porter, L.W.; Steele, L.P.; Trudinger, C.M.; et al. Perfluorocarbons in the global atmosphere: Tetrafluoromethane, hexafluoroethane, and octafluoropropane. *Atmos. Chem. Phys.* **2010**, *10*, 5145–5164. [CrossRef]
9. Corr, S. 1,1,1,2-Tetrafluoroethane; from refrigerant and propellant to solvent. *J. Fluor. Chem.* **2002**, *118*, 55–67. [CrossRef]
10. Akasaka, R.; Zhou, Y.; Lemmon, E.W. A Fundamental Equation of State for 1,1,1,3,3-Pentafluoropropane (R-245fa). *J. Phys. Chem. Ref. Data* **2015**, *44*, 013104. [CrossRef]
11. The European Parliament and the Council of the European Union. Regulation (EU) No 517/2014 of the European Parliament and of the Council of 16 April 2014 on Fluorinated Greenhouse Gases and Repealing Regulation (EC) No 842/2006. 2014, 150/195-230. Available online: https://eur-lex.europa.eu/eli/reg/2014/517/oj (accessed on 14 December 2021).
12. European Commission. Climate Action: EU Legislation to Control F-Gases. Available online: https://ec.europa.eu/clima/policies/f-gas/legislation_en (accessed on 14 December 2021).

13. Nagarsheth, R.; Singh, S. Study of gas insulated substation and its comparison with air insulated substation. *Int. J. Electr. Power Energy Syst.* **2014**, *55*, 481–485. [CrossRef]

14. Sato, J.; Sakaguchi, O.; Kubota, N.; Makishima, S.; Kinoshita, S.; Shioiri, T.; Yoshida, T.; Miyagawa, M.; Homma, M.; Kaneko, E. New technology for medium voltage solid insulated switchgear. In Proceedings of the IEEE Transmission and Distribution Conference and Exhibition, Yokohama, Japan, 6–10 October 2002.

15. Hyrenbach, M.; Zache, S. Alternative insulation gas for medium-voltage switchgear. In Proceedings of the 2016 Petroleum and Chemical Industry Conference Europe, Berlin, Germany, 14–16 June 2016.

16. Kieffel, Y.; Irwin, T.; Ponchon, P.; Owens, J. Green Gas to Replace SF_6 in Electrical Grids. *IEEE Power Energy Mag.* **2016**, *14*, 32–39. [CrossRef]

17. Pionteck, J.; Wypych, G. *Handbook of Antistatics*, 2nd ed.; ChemTech Publishing: Toronto, ON, Canada, 2016.

18. Kieffel, Y.; Biquez, F. SF6 alternative development for high voltage switchgears. In Proceedings of the 2015 IEEE Electrical Insulation Conference, Seattle, WA, USA, 7–10 June 2015.

19. Izcara, J.; Larrieta, J. Electrical Insulation System for Medium- and High-voltage Electrical Switchgear. U.S. Patent 10,607,748 B2, 31 March 2020.

20. Glasmacher, P. Fluorinated Ketones as a High-Voltage Insulating Medium. WO Patent 2010/146022A1, 23 December 2010.

21. Mantilla, J.; Claessens, M.-S.; Gariboldi, N.; Grob, S.; Skarby, P.; Paul, T.A.; Mahdizabed, N. Dielectric Insulation Medium. U.S. Patent 8,822,870 B2, 2 September 2014.

22. Kieffel, Y.; Girodet, A.; Piccoz, D.; Maladen, R. Use of a Mixture Comprising a Hydrofluoolefin as a Medium-voltage Arc-extiguishing and/or Insulating Gas and Medium-Voltage Electrical Device Comprising Same. U.S. Patent 9,491,877, 8 November 2016.

23. Preve, C.; Piccoz, D.; Maladen, R. Validation methods of SF_6 alternatives gas. In Proceedings of the 23rd International Conference on Electricity Distribution, Lyon, France, 15–18 June 2015.

24. Pang, X.; Wu, H.; Pan, J.; Qi, Y.; Li, X.; Zhang, J.; Xie, Q. Analysis of correlation between Internal discharge in GIS and SF_6 decomposition products. In Proceedings of the IEEE International Conference on High Voltage Engineering and Application, Athens, Greece, 10–13 September 2018.

25. Katagiri, H.; Kasuya, H.; Mizoguchi, H.; Yanabu, S. Investigation of the performance of CF_3I Gas as a Possible Substitute for SF6. *IEEE Trans. Dielectr. Electr. Insul.* **2008**, *15*, 1424–1429. [CrossRef]

26. Skaggs, S.R.; Rubenstein, R. Setting the occupational exposure limit for CF3I. In Proceedings of the Halon Options Technical Working Conference, Albuquerque, NM, USA, 27–29 April 1999.

27. Widger, P.; Haddad, A. Analysis of gaseous by-products of CF_3I and CF_3I-CO_2 after high voltage arcing using a GCMS. *Molecules* **2019**, *24*, 1599. [CrossRef]

28. Wang, Y.; Gao, Z.; Wang, B.; Zhou, W.; Yu, P.; Luo, Y. Synthesis and dielectric properties of trifluoromethanesulfonyl fluoride: An alternative gas to SF_6. *Ind. Eng. Chem. Res.* **2019**, *58*, 21913–21920. [CrossRef]

29. Hu, S.; Wang, Y.; Zhou, W.; Qiu, R.; Luo, Y.; Wang, B. Dielectric Properties of CF_3SO_2F/N_2 and CF_3SO_2F/CO_2 Mixtures as a Substitute to SF6. *Ind. Eng. Chem. Res.* **2020**, *59*, 15796–15804. [CrossRef]

30. Long, Y.; Guo, L.; Wang, Y.; Chen, C.; Chen, Y.; Li, F.; Zhou, W. Electron Swarms Parameters in CF_3SO_2F as an Alternative Gas to SF6. *Ind. Eng. Chem. Res.* **2020**, *59*, 11355–11358. [CrossRef]

31. Wang, X.; Zhong, L.; Yan, J.; Yang, A.; Han, G.; Wu, Y.; Rong, M. Investigation of dielectric properties of cold C_3F_8 mixtures and hot C3F8 gas as Substitutes for SF_6. *Eur. Phys. J. D* **2015**, *69*, 240. [CrossRef]

32. Deng, Y.; Li, B.; Xiao, D. Analysis of the insulation characteristics of C_3F_8 gas mixtures with N_2 and CO_2 using boltzmann equation method. *IEEE Trans. Dielectr. Electr. Insul.* **2015**, *22*, 3253–3259. [CrossRef]

33. Trudinger, C.M.; Fraser, P.J.; Etheridge, D.M.; Sturges, W.T.; Vollmer, M.K.; Rigby, M.; Martinerie, P.; Mühle, J.; Worton, D.R.; Krummel, P.B.; et al. Atmospheric abundance and global emissions of perfluorocarbons CF_4, C_2F_6 and C_3F_8 since 1800 inferred from ice core, firn, air archive and in situ measurements. *Atmos. Chem. Phys.* **2016**, *16*, 11733–11754. [CrossRef]

34. Li, Y.; Zhang, X.; Xiao, S.; Chen, Q.; Tang, J.; Chen, D.; Wang, D. Decomposition Properties of C_4F_7N/N_2 Gas Mixture: An Environmentally Friendly Gas to Replace SF_6. *Ind. Eng. Chem. Res.* **2018**, *57*, 5173–5182. [CrossRef]

35. Chachereau, A.; Hösl, A.; Franck, C.M. Electrical insulation properties of the perfluoronitrile C_4F_7N. *J. Phys. D Appl. Phys.* **2018**, *51*, 495201. [CrossRef]

36. Andersen, M.P.S.; Kyte, M.; Andersen, S.T.; Nielsen, C.J.; Nielsen, O.J. Atmospheric Chemistry of $(CF_3)_2CF–C\equiv N$: A Replacement Compound for the Most Potent Industrial Greenhouse Gas, SF_6. *Environ. Sci. Technol.* **2016**, *51*, 1321–1329. [CrossRef]

37. Zhao, M.; Han, D.; Zhou, Z.; Zhang, G. Experimental and theoretical analysis on decomposition and by-product formation process of $(CF_3)_2CFCN$ mixture. *AIP Adv.* **2019**, *9*, 105204. [CrossRef]

38. Koch, M.; Franck, C.M. High voltage insulation properties of HFO1234ze. *IEEE Trans. Dielectr. Electr. Insul.* **2015**, *22*, 3260–3268. [CrossRef]

39. Liang, Y.; Wang, F.; Sun, Q.; Zhai, Y. Molecular structural and electrical properties of trans-1,3,3,3-tetrafluoropropene and 2,3,3,3-tetrafluoropropene under external electric fields. *Comput. Theor. Chem.* **2017**, *1120*, 79–83. [CrossRef]

40. Wang, J.; Li, Q.; Liu, H.; Huang, X.; Wang, J. Theoretical and experimental investigation on decomposition mechanism of eco-friendly insulation gas HFO1234zeE. *J. Mol. Graph. Model.* **2020**, *100*, 107671. [CrossRef] [PubMed]

41. Mantilla, J.D.; Gariboldi, N.; Grob, S.; Claessens, M. Investigation of the insulation performance of a new gas mixture with extremely low GWP. In Proceedings of the IEEE Electrical Insulation Conference, Philadelphia, PA, USA, 8–11 June 2014.

42. Simka, P.; Ranjan, N. Dielectric strength of C5 perfluoroketone. In Proceedings of the 19th International Symposium on High Voltage Engineering, Pilsen, Czech Republic, 23–28 August 2015.

43. Li, Y.; Zhang, X.; Xiao, S.; Chen, Q.; Wang, D. Decomposition characteristics of $C_5F_{10}O$/air mixture as substitutes for SF_6 to reduce global warming. *J. Fluor. Chem.* **2018**, *208*, 65–72. [CrossRef]

44. Preve, C.; Piccoz, D.; France, P.S.; Maladen, R. Comparison of Alternatives to SF6 Regarding EHS and End of Life. In Proceedings of the 25th International Conference on Electricity Distribution, Madrid, Spain, 3–6 June 2019.

45. Zhuo, R.; Chen, Q.; Wang, D.; Fu, M.; Tang, J.; Hu, J.; Jiang, Y. Compatibility between C6F12O–N2 Gas Mixture and Metal Used in Medium-Voltage Switchgears. *Energies* **2019**, *12*, 4639. [CrossRef]

46. Blázquez, S.; Antiñolo, M.; Nielsen, O.J.; Albaladejo, J.; Jiménez, E. Reaction kinetics of $(CF_3)_2CFCN$ with OH radicals as a function of temperature (278–358 K): A good replacement for greenhouse SF_6? *Chem. Phys. Lett.* **2017**, *687*, 297–302. [CrossRef]

47. Kramer, A.; Over, D.; Stoller, P.; Paul, T.A. Fiber-coupled LED gas sensor and its application to online monitoring of ecoefficient dielectric insulation gases in high-voltage circuit breakers. *Appl. Opt.* **2017**, *56*, 4505. [CrossRef]

48. Saxegaard, M.; Seeger, M.; Kristoffersen, M.; Germany, A.; Stoller, P.; Landsverk, H. Dielectric properties of gases suitable for secondary medium voltage switchgear. In Proceedings of the 23rd International Conference om Electricity Distribution, Lyon, France, 15–18 June 2015.

49. Kuffel, E.; Zaengl, W.S.; Kuffel, J. *High Voltage Engigeering Fundamentals*, 2nd ed.; Elsevier: Amsterdam, The Netherlands, 2000.

50. Maikala, R.V. Modified Beer's Law—Historical perspectives and relevance in near-infrared monitoring of optical properties of human tissue. *Int. J. Ind. Ergon.* **2010**, *40*, 125–134. [CrossRef]

51. Seader, J.D.; Henley, E.J.; Roper, D.K. *Separation Process Principles*; Wiley: Hoboken, NJ, USA, 2011.

52. Taylor, R.; Krishna, R. *Multicomponent Mass Transfer*; Wiley: Hoboken, NJ, USA, 1993.

53. Bird, R.B.; Stewart, W.E.; Lightfoot, E.N. *Transport Phenomena*, 2nd ed.; Wiley: Hoboken, NJ, USA, 2002.

54. Johnson, R.A. *Miller and Freud's Probability and Statistics for Engineers*, 9th ed.; Pearson: London, UK, 2018.

55. Fuller, E.N.; Schettler, P.D.; Giddings, J.C. A new method for prediction of binary gas-phase diffusion coefficients. *Ind. Eng. Chem.* **1966**, *58*, 18–27. [CrossRef]

56. Fuller, E.N.; Ensley, K.; Giddings, J.C. Diffusion of halogenated hydrocarbons in helium. The effect of structure on collision cross sections. *J. Phys. Chem.* **1969**, *73*, 3679–3685. [CrossRef]

57. Hu, X.; Yu, X.; Hou, H.; Wang, B. Stereo-dependent dimerization, boiling points, diffusion coefficients, and dielectric constants of E/Z-HFO-1234ze. *Int. J. Quantum Chem.* **2022**, *122*, e26848. [CrossRef]

Article

The Choice of Optical Flame Detectors for Automatic Explosion Containment Systems Based on the Results of Explosion Radiation Analysis of Methane- and Dust-Air Mixtures

Sergey Khokhlov *, Zaur Abiev and Viacheslav Makkoev

Department of Blasting, Saint Petersburg Mining University, 199106 Saint Petersburg, Russia;
abiev_za@pers.spmi.ru (Z.A.); viatcheslav.makkoev@gmail.com (V.M.)
* Correspondence: khokhlov_sv@pers.spmi.ru

Abstract: A review of the existing optoelectron monitoring devices revealed that the design of optoelectron detectors of the mine atmosphere does not sufficiently take into account the factor of external optical interference. This includes any extraneous source of thermal emission: a source of artificial lighting or enterprises. As a consequence, the optoelectron detectors -based safety systems currently installed at mining sites are not able to ensure properly the detection of the ignition source in the presence of optical interference. Thus, it is necessary to determine the working spectral wavelength ranges from methane and coal dust explosions. The article presents the results of experimental research devoted to the methane-air mixture and coal dust explosion spectral analysis by means of the photoelectric method. The ignition of a methane-air mixture of stoichiometric concentration (9.5%) and coal dust of size characterized by the dispersion of 63–94 microns and concentration of 200 g/m^3 was carried out in a 20 L spherical chamber with an initial temperature in the setup of 18–22 °C at atmospheric pressure. Then, photometry of the explosion light flux was conducted on a photoelectric unit. Operating spectral wavelength ranges from methane and coal dust explosions were determined. For the methane-air mixture, it is advisable to use the spectral regions at the maximum emission of 390 and 900 nm. The spectrum section at the maximum emission of 620 nm was sufficient for dust-air mixture. It enabled us to select the wavelength ranges for automatic explosion suppression systems' launching references. This will exclude false triggering of the explosion suppression system from other radiation sources. The research results will help to improve the decision-making credibility of the device in its direct design. The results will be used in further research to design noise-resistant optical flame detection sensors with a high response rate.

Keywords: explosion; coal dust; methane; explosion suppression; spectral characteristics; explosion pressure; radiation intensity; free radicals

Citation: Khokhlov, S.; Abiev, Z.; Makkoev, V. The Choice of Optical Flame Detectors for Automatic Explosion Containment Systems Based on the Results of Explosion Radiation Analysis of Methane- and Dust-Air Mixtures. *Appl. Sci.* **2022**, *12*, 1515. https://doi.org/10.3390/app12031515

Academic Editor: Andreas Fischer

Received: 27 December 2021
Accepted: 28 January 2022
Published: 30 January 2022

Publisher's Note: MDPI stays neutral with regard to jurisdictional claims in published maps and institutional affiliations.

Copyright: © 2022 by the authors. Licensee MDPI, Basel, Switzerland. This article is an open access article distributed under the terms and conditions of the Creative Commons Attribution (CC BY) license (https://creativecommons.org/licenses/by/4.0/).

1. Introduction

Dust and gas explosions are among the greatest disasters in the coal industry and are related with mass fatalities. This is a danger for the entire mining industry, not just coal mining [1]. In this regard, safety improvement during blasting operations in gaseous and dusty mines is of utmost importance [2]. Dust-methane explosion safety upgrades are possible only by a comprehensive approach, including risk management [3–5], development of tools and methods of mine explosion protection, methane emission control [6], coal deposits underground mining technological processes monitoring [7–10].

Among the effective ways to control possible explosions are the automatic ignition prevention systems. The sensors applied in automatic explosion barriers respond to high pressure (outside rods) and temperature, abnormal concentrations of explosive gases, optical parameters and flames [11]. The disadvantage of the sensors, responding to the explosion wave pressure, is their possible actuation because of extraneous acoustic signals. The authors of [12] provide an explanation of the outside rods' inefficiency due to their spatiotemporal parameters that are insufficient for successful explosion containment. The disadvantage of sensors, responding to the temperature changes, is their sensibility to the dusty mine environment, dust deposits, etc. [13]. Since the ignition process includes lighting, the perfect solution is optical sensors detecting dynamics in infrared and ultraviolet radiation and able to start the automatic prevention system to avoid combustion and explosions of methane-dust-air mixture.

The first thing we consider when estimating occupational injury risks is the air shock wave produced by methane-dust-air mixture's explosion [9]. Therefore, it's clear that an effective suppression system should be started at the very early ignition stages [14–16]. For this reason, the optical sensor's spectral response should be sufficient to start the suppression system.

There is one method [17,18] to measure object temperature without known emittance involvement. However, this method is not effective because of its slow response, due to the need to analyze and process a wide range of the optical spectrum [19]. In this regard, the correct choice of optical sensor is impossible without considering a specific flame and methane-dust-air explosion radiation spectrum.

A number of optoelectron devices (OEDs) have been proposed for use in fire detection and localization systems, including two spectral ratio optoelectron devices. The device uses radiation in three spectral ranges (750 ± 40 nm, 950 ± 50 nm, 1550 ± 12 nm). However, the study does not give a criterion for the choice of these ranges [20].

Moreover, for proper extinguishment within the development ratio with a cross-sectional area up to 10 m^2, it is necessary to create the explosion-suppressing environment by throwing out at least 30 kg of inhibitor within 15 ms. This throwing rate in case of false alarms can be dangerous for people located in the immediate vicinity to the explosion suppression devices [21].

Therefore, the development of construction and creation of a fast-operating OED control of explosive dust and gas atmosphere, insensitive to dustiness of the intermediate atmosphere and having a high probability of detecting the ignition source at an early stage in the presence of external optical interference, is an urgent scientific and technological task. It has an essential economic importance.

Thus, this research aims to obtain the spectral characteristics of combustion and methane-air mixtures' explosion radiation for the correct choice of input sensors applicable for explosion suppression systems.

2. Materials and Methods

The research into the flame radiation spectrum and methane, air and dust mixtures' explosion radiation spectrum was carried out by photoelectric method and aimed to obtain data on the nature of radiation energy wavelength distribution. This method enables spectrum recording with automatic dark signal subtraction and spectrograph wavelength calibration [22,23].

Currently it has been established [24,25] that gas radiation, heated by a shock wave, corresponds to "gray body" radiation, provided that the relative spectral energy distribution can be almost identified with any "black body" energy distribution at the temperature T.

The spectral-energy distribution of radiation emitted by a "black body" is described by the Planck formula:

$$B_\lambda = \frac{C_1 \cdot \lambda^{-5}}{e^{C_2/\lambda \cdot T} - 1},$$ (1)

where B_λ is emitting surface spectral brightness, λ is emission wavelength, C_1 and C_2 are constants, and T is emitting surface temperature.

Provided that $C_2 / \lambda \cdot T \gg 1$, Formula (1) is reduced to the classic Wien formula, which gives quite an accurate description of shock wave spectral energy distribution for the visible area:

$$B_\lambda = c_1 \lambda^{-5} e^{-\frac{c_2}{\lambda T}}.$$ (2)

Formula (2) shows that for the curve reflecting spectral energy distribution of the luminous gases, which are gray emitters, it is necessary to provide its color temperature measurements.

The method of color temperature measurement is based on light intensity comparison from two spectrum areas.

Using Equation (2) we can have the following formula for energy ΔE, emitting by the body at the temperature T, wavelength λ and in the bandwidth $\Delta \lambda$:

$$\Delta E_{\lambda,T} = c_1 \lambda^{-5} \left(e^{-\frac{c_2}{\lambda T}} \right) \Delta \lambda.$$ (3)

If we know $\Delta E_{\lambda,T}$ for different wavelengths λ_1 and λ_2, it is easy to calculate the emitter temperature, when the reference source and temperature T_x are available.

We have:

$$\Delta E_{1x} = \Delta E_{\lambda_1,T_x} = C_2 \lambda_1^{-5} e^{-\frac{C_2}{\lambda_1 T_x}} \Delta \lambda_1,$$ (4)

$$\Delta E_{2x} = \Delta E_{\lambda_2,T_x} = C_2 \lambda_2^{-5} e^{-\frac{C_2}{\lambda_2 T_x}} \Delta \lambda_2.$$ (5)

Dividing Equation (3) by Equation (4) and taking logarithms, we have the following expression for the emitter under study:

$$\ln \left(\frac{\Delta E_{1x}}{\Delta E_{2x}} \right) = -5 \ln \left(\frac{\lambda_1}{\lambda_2} \right) + \ln \left(\frac{\Delta \lambda_1}{\Delta \lambda_2} \right) - \frac{C_2}{T_x} \left(\frac{1}{\lambda_1} - \frac{1}{\lambda_2} \right)$$ (6)

and for the reference source:

$$\ln \left(\frac{\Delta E_{10}}{\Delta E_{20}} \right) = -5 \ln \left(\frac{\lambda_1}{\lambda_2} \right) + \ln \left(\frac{\Delta \lambda_1}{\Delta \lambda_2} \right) - \frac{C_2}{T_0} \left(\frac{1}{\lambda_1} - \frac{1}{\lambda_2} \right).$$ (7)

From Equations (6) and (7) we have the equation for the source color temperature T_x calculation using the available temperature value of the reference emitter T_0:

$$\ln \left(\frac{\Delta E_{10}/\Delta E_{20}}{\Delta E_{1x}/\Delta E_{2x}} \right) = C_2 \left(\frac{1}{\lambda_1} - \frac{1}{\lambda_2} \right) \cdot \left(\frac{1}{T_x} - \frac{1}{T_0} \right).$$ (8)

A general algorithm of the laboratory experiment is the following: in a closed combustion chamber of a 20-L spherical explosion chamber [26–29] shown in Figure 1, we have a mixture of stoichiometric concentration, supplied by a single step by compression with 2 MPa pressure [30,31]. The accidents in coal mines mainly occur because of methane and coal dust explosions [1,32]. It is these two components that were the object of the study. Further, it is flamed with 60 ms delay. The initial temperature in the unit is about 18–22 °C at the atmospheric pressure. The tested samples were prepared using the partial pressure method. Then, before ignition, the mixture was stirred by circulation pump, to ensure its homogeneity [33]. A luminous flow produced by the combustible mixture ignition was observed through the explosion chamber watch window.

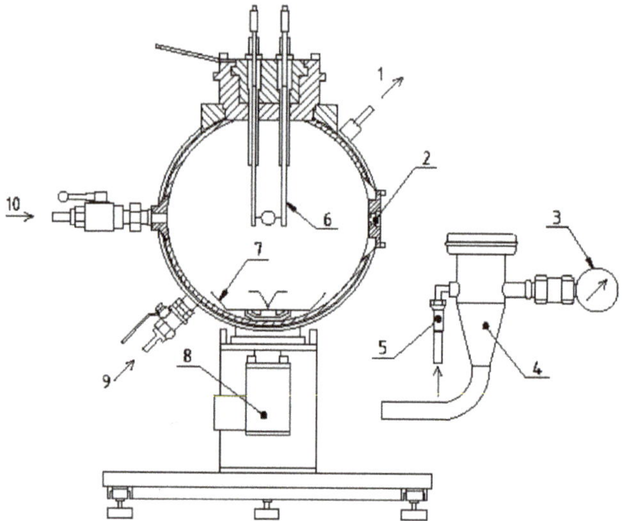

Figure 1. Twenty-liter spherical explosion chamber. The unit layout. 1—Water output, 2—pressure sensor, 3—pressure gauge, 4—dust collector 0.6 dm³, 5—air intake, ignition source, 6—chemical igniters, 7—rebound sprayer, 8—fast-acting valve, 9—water intake, 10—air and resultant outlet.

The results of aerosol ignition of certain concentration, that was produced inside the chamber, as well as the explosion pressure and the explosion pressure rise rate are automatically recorded by the data processing system. This further data analysis enables us to conclude which is the successful ignition mixture (Figure 2).

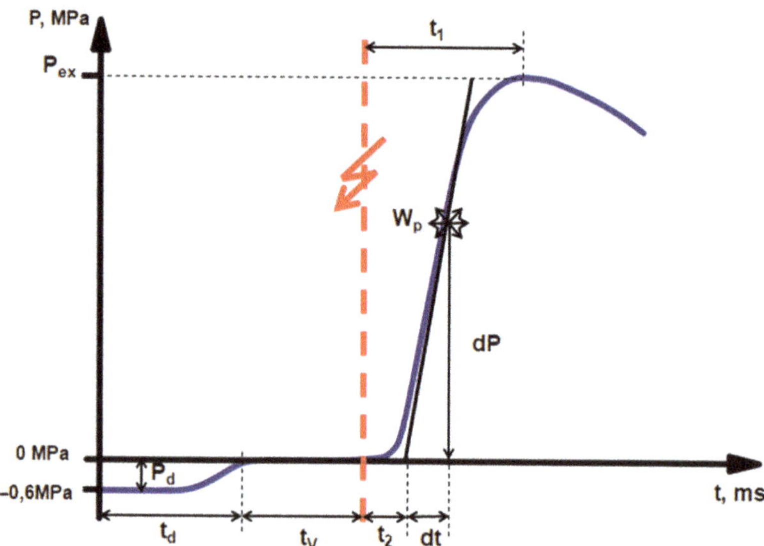

Figure 2. Pressure trend (P, MPa) for the period (t, ms) of dust-gas mixture combustion inside the explosion chamber: P_d—expansion pressure of the combustion chamber; P_{ex}—explosion pressure; td—exhaust valve delay; t_1—combustion time; t_2—induction time; t_v—ignition delay time; W_p—breakpoint in the rising part of the pressure curve; dP/dt—pressure rise rate at the explosion.

A pressure gauge with a response time of 0.2 ms is applicable for pressure measurements up to 2 MPa. Timing of pressure and flame radiation spectrum data recording from the ignition moment was controlled by ExTest software. To exclude the influence of the decomposition products of chemical igniters on the test result, the mixture was ignited by flash over. The energy produced from electrical initiation was 1 kJ.

Photometric measurement of light fluxes involved in the Equation (8) was carried out by means of photoelectric unit (Figure 3).

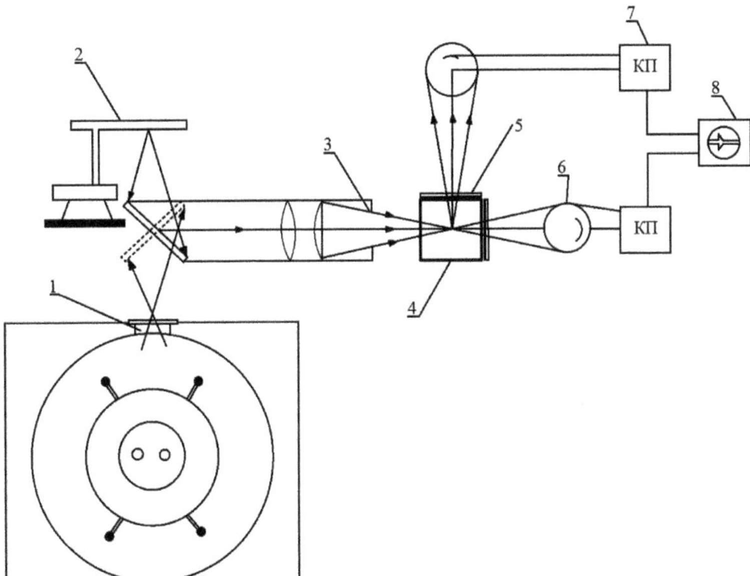

Figure 3. Basic diagram of temperature measurement. 1—Explosion chamber watch window, 2—band-lamp, 3—diaphragm, 4—splitter, 5—light filter, 6—photo multiplier, 7—cathode amplifier, 8—oscillograph.

The luminous flux was projected onto the inlet diaphragm of the unit using a rotatable flat mirror and lens. The luminous flux, after passing through the diaphragm, was recorded by photomultipliers using a beam splitting system, the signals of which were recorded from the oscilloscope screen. Two spectral intervals were separated by means of 15 nm bandpass interference filters.

The recording device was calibrated using standard stripe incandescent lamp. Its luminous flux was projected onto the inlet diaphragm of the unit through a hole in a rotating disk (chopper) using a rotary flat mirror and lens.

To measure the absolute radiation intensity, we used photomultipliers powered by high-voltage rectifier with electronic regulation.

For the radiator processes recording, it is necessary to consider not just the receiver's absolute sensitivity, but also the wavelength interval where this sensitivity remains effective.

One of the basic parameters of photodetectors is their time constant of the order of 10^{-8}–10^{-10} s.

The photomultipliers are characterized by significant photocurrent amplification factor and are well protected from the interferences caused by external electric fields. A relatively high input current of multistage photomultiplier enables to record output signals by cathode oscilloscopes without special broadband amplifiers involvement.

The oscilloscope's beam deviations are proportional to the radiation energy of the selected spectral intervals

$$D = \kappa \Delta E_1 \tag{9}$$

where D is oscilloscope beam deviation, and ΔE is defined by Formula (3). Using Equation (8) we have:

$$\ln\left(\frac{D_{10}/D_{20}}{D_{1x}/D_{2x}}\right) = C_2\left(\frac{1}{\lambda_1} - \frac{1}{\lambda_2}\right) \cdot \left(\frac{1}{T_x} - \frac{1}{T_0}\right) \tag{10}$$

Let us denote the glow signals ratio of the investigated medium for the selected spectrum ranges by α, and the ratio of the calibration signals from a reference source at special color temperature T by β:

$$\alpha = D_{1x}/D_{2x} \tag{11}$$

$$\beta = D_{10}/D_{20} \tag{12}$$

If we substitute Equation (11) and Equation (12) into Equation (10) we'll get a working formula for source temperature calculation:

$$\ln(\beta/\alpha) = c_2\left(\frac{1}{\lambda_1} - \frac{1}{\lambda_2}\right) \cdot \left(\frac{1}{T_x} - \frac{1}{T_o}\right) \tag{13}$$

If the reference source temperature is well known, then the temperature measurement error is calculated by the following formula:

$$\frac{\Delta T_x}{T_x} = \frac{\lambda_1 \cdot \lambda_2}{c_2(\lambda_2 - \lambda_1)} \cdot \left(\frac{\Delta \alpha}{\alpha} + \frac{\Delta \beta}{\beta}\right) \tag{14}$$

The measurement error is mostly because of the finite thickness of the oscilloscope beam and photoelectronic multiplier noises. According to the calculations, the relative error does not exceed 10%.

3. Results

As shown in Figure 4a,b the mixture explosiveness assessment was carried out considering pressure values P_{max} and the explosion pressure rise rate $(dP/dt)_{max}$ [34]. To show the spectral radiation characteristics, the intensity (I) was recorded and analyzed as it is shown in Figure 5. Parameter I_0 is defined as the value at which the radiation intensity deviates from the baseline (I_0 is 110% of the initial radiation intensity) [35]. The parameter t0 shows the time necessary to reach the maximum value of radiation intensity.

According to the received experimental data, the ignited methane of stoichiometric concentration in chamber produced maximal explosion pressure $P_{max} = 0.83$ MPa and the rate of pressure rise was 42.05 MPa/s. When the explosive combustion of KS (KC) grade coal dust was registered in the Dzerzhinsky mine site, the maximum explosion pressure in the chamber was 0.79 MPa and the pressure rise rate was 34.62 MPa/s.

The results were analyzed using the application software. According to the obtained data, the graphs of changes in the pressure of methane and coal dust explosion were plotted against the time of explosive combustion of the mixture.

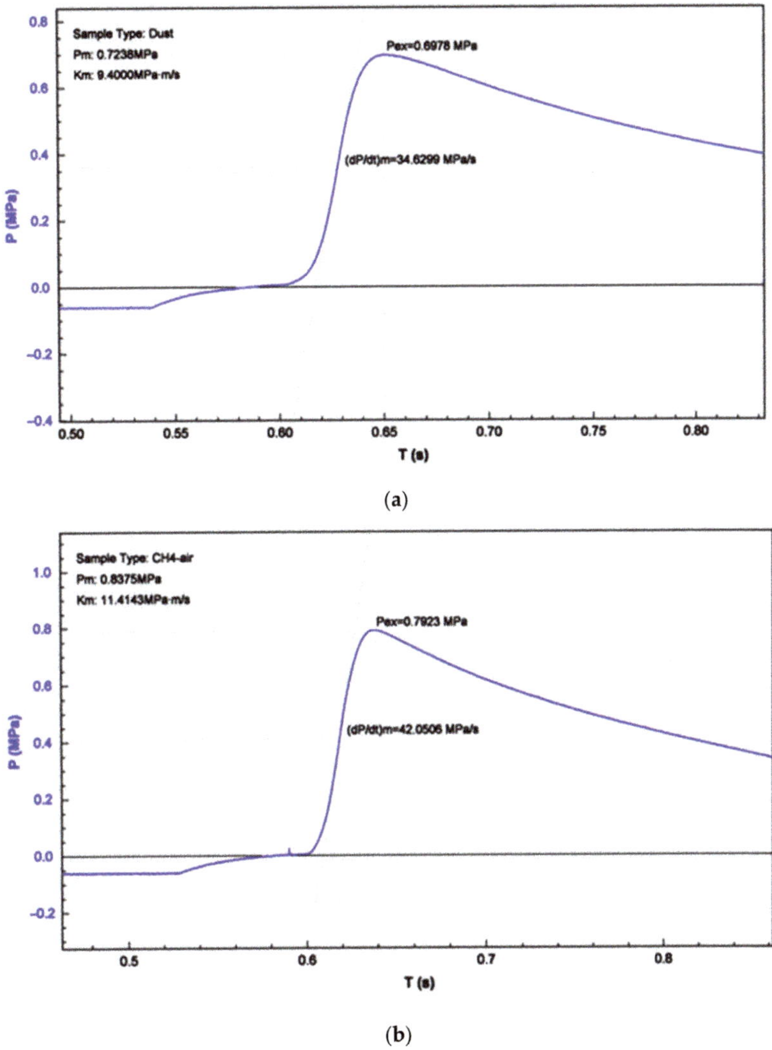

Figure 4. (**a**)Pressure-time curve based on CH_4 ignition (9.5% Vol.). (**b**) Pressure-time curve based on the results of the coal dust ignition (concentration 200 g/m^3, dispersibility 63–94 microns).

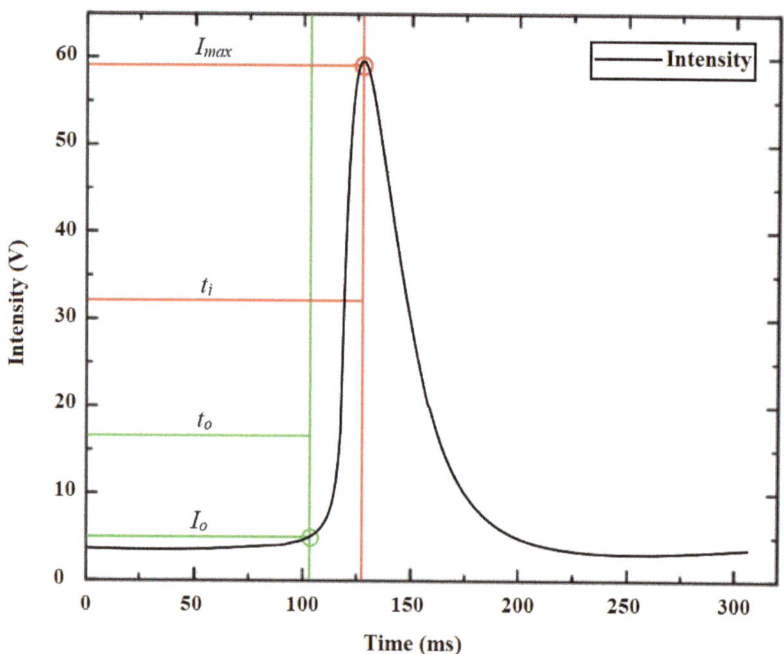

Figure 5. Maximum intensity of spectral radiation against time.

The results of experimental studies of radiation intensity measurement of methane-air and dust-air mixture explosion carried out by the electron-optical method are presented in the form of a dependence diagram, Figure 6.

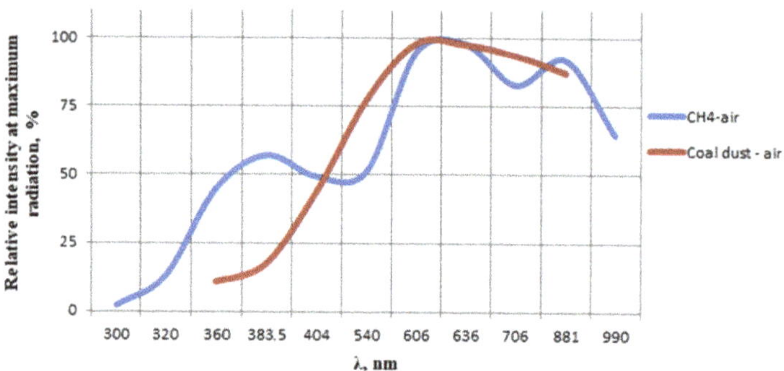

Figure 6. Spectral energy distribution at maximum radiation of methane-dust-air mixtures.

The relative intensity dependence on the wavelength shows that wavelength distribution of radiation energy differs from the Planck's type of distribution. The graph (Figure 6) clearly shows two peaks at the wavelengths $\lambda_1 = 383.5$ nm and $\lambda_2 = 620$ nm. These peaks appeared when the banded spectrum of molecules and radicals overlaid the continuous radiation spectrum of the heated gas.

The peak point in the range of $\lambda = 390$ nm can be identified with band system of the radical CH radiation [36], and the one in the range of $\lambda = 620$ nm–with the band system of the molecule C_2 radiation (Swan system). The band system of molecule C_2 is also in the range of $\lambda_1 = 380$ nm.

An intense series of bands are observed in the methane combustion spectrum, caused by the OH radical emission in the ultraviolet range of the spectrum $\lambda_1 = 306$ nm. The maximum radiation in the range of 900 nm should be identified with methane-air continuous combustion spectrum. In this case, the temperature, that was determined by the curve peak in accordance with Wien's displacement law, is $T = 2610$ K, which correlates well with the data received by spectrometric methods [37].

4. Conclusions

The availability of free radicals, which are considered the active centers of the chain reaction of methane explosion, is necessary for the entire process of explosive transition. Some free radicals, such as OH, H, O, CH_3, HO_2, CHO, are especially important not only for the explosion process, but also for the explosion suppression (inhibition) [38–40].

The flame spectrum analysis reveals the intermediate compounds formed during combustion and explosion and lets us study their behavior. Nowadays optical spectroscopy is the best method for free radical detection, since this method has no impact on the combustion and explosion process. The comparison of spectrum relative intensity enables us to get data on chemical reactions and involved radicals.

The resulting dependences of energy spectral distribution of explosions of methane-dust-air mixtures enable us to determine the central wavelengths of bandpass filters and select the spectral ranges for input sensors and explosion suppression.

Thus, for methane-air mixtures, it is reasonable to use spectral regions at the radiation maximums of 390 nm and 900 nm. This helps to avoid false triggering of the explosion suppression system possibly initiated by other radiation sources. For a dust-air mixture, it is enough to use one spectral region at the radiation maximum of 620 nm.

Thus, it is proved that for the development of an active explosion suppression system in coal mines, particular attention should be paid to the choice of working spectral wavelength ranges for the recognition of the desired signal.

The most important parameter for optoelectron devices is the credibility of the decision. It is a complex parameter and is determined by a combination of the following:

- Probability of fire detection in the absence of optical interference;
- False alarm probability;
- Probability of fire detection in the presence of external optical interference. The research results will help to improve the decision-making credibility of the device in its direct design.

We believe that further research should be devoted to initial combustion detection technology development with its further integration into multifunctional safety systems purposed for successful methane-dust-air combustion and explosion containment in coal mines and for the mines' industrial testing safety.

Author Contributions: Development of methodology, Z.A. and S.K.; creation of models, Z.A.; preparation and writing of the initial version, Z.A. and V.M.; provision of research materials, Z.A.; general control over the work, S.K.; mathematical processing, V.M.; data representation, V.M. All authors have read and agreed to the published version of the manuscript.

Funding: The study was carried out at the expense of a subsidy for the fulfillment of the state task in the field of scientific activity for 2021 No. FSRW-2020-0014.

Institutional Review Board Statement: Not applicable.

Informed Consent Statement: Not applicable.

Conflicts of Interest: The authors declare no conflict of interest.

References

1. Goncharov, E.V.; Tsirel', S.V. Geodynamic methods for assessing methane distribution in bituminous coal deposits and measures to intensify methane fluxes during mine gas drainage. *J. Min. Inst.* **2016**, *222*, 803. [CrossRef]

2.	Rodionov, V.A.; Pikhonen, L.V.; Zhiharev, S.Y. Analyzing application methods of thermal analysis for evaluation of explosive properties of Sokolovsky deposit coal. *Proc. Tula States Univ.-Sci. Earth* **2017**, *3*, 84–93.

3.	Yutiaev, E.; Meshkov, A.; Popov, A.; Shabarov, A. Allocation of the geo-dynamically hazardous zones during intensive mining of flat-lying coal seams in the mines of SUEK-Kuzbass JSC. *E3S Web Conf.* **2019**, *134*, 01022. [CrossRef]

4.	Golubev, D.D. Influence of technological factors on the formation of spontaneous combustion centers in underground mining. In *Scientific and Practical Studies of Raw Material Issues*; CRC Press: Boca Raton, FL, USA, 2019; pp. 75–81. [CrossRef]

5.	Strizhenok, A.V.; Ivanov, A.V. An Advanced Technology for Stabilizing Dust Producing Surfaces of Built-Up Technogenic Massifs during Their Operation. *Power Technol. Eng.* **2016**, *50*, 240–243. [CrossRef]

6.	Meshkov, A.; Kazanin, O.; Sidorenko, A. Methane Emission Control at the High-Productive Longwall Panels of the Yalevsky Coal Mine. *E3S Web Conf.* **2020**, *174*, 01040. [CrossRef]

7.	Dmitrievich, M.R.; Alekseevich, R.V.; Borisovich, S.V. Methodological Approach to Issue of Researching Dust-Explosion Protection of Mine Workings of Coal Mines. *Int. J. Civ. Eng. Technol.* **2019**, *10*, 1154–1161.

8.	Kabanov, E.I.; Korshunov, G.I.; Magomet, R.D. Quantitative risk assessment of miners injury during explosions of methane-dust-air mixtures in underground workings. *J. Appl. Sci. Eng.* **2020**, *24*, 105–110.

9.	Smirnyakov, V.; Smirnyakova, V.; Pekarchuk, D.; Orlov, F. Analysis of methane and dust explosions in modern coal mines in Russia. *Int. J. Civ. Eng. Technol.* **2019**, *10*, 1917–1929.

10.	Koteleva, N.; Kuznetsov, V.; Vasilyeva, N. A Simulator for Educating the Digital Technologies Skills in Industry. Part One. Dynamic Simulation of Technological Processes. *Appl. Sci.* **2021**, *11*, 10885. [CrossRef]

11.	Shevtsov, N. Vzryvozaschita gornyh vyrabotok: Uchebnoe posobie dlya vuzov. In *Explosion Mining: A Manual for Schools*, 2nd ed.; Nord-Press: Donetsk, Russian, 2002.

12.	Vasil'ev, A.A. Estimates of the excitation conditions and extinction of blast waves during mine explosions. *Bull. Res. Cent. Saf. Coal Ind.* **2016**, *2*, 91–105.

13.	Necepljaev, M.I. Bor'ba so vzryvami ugol'noj pyli v shahtah. In *Fighting Coal Dust Explosions in Mines*; Nedra: Moscow, Russian, 1992; p. 298.

14.	Sidorenko, A.I. Pyrometric sensor with optical shutters for determining the two-dimensional coordinates of the explosion center. *Vestn. Nauchnogo Cent. be-Zopasnosti Rabot Ugol'noj Promyshlennosti* **2013**, *1*, 98–104.

15.	Ajruni, A.T.; Klebanov, F.S.; Smirnov, O.V. Vzryvoopasnost' ugol'nyh shaht. In *Explosive-Ness of Coal Mines*; Gornoe delo OOO Kimmerijskij Centr: Moscow, Russian, 2011; p. 264.

16.	Zaharenko, D.M. Problemy rannego obnaruzhenija ochagov pozharov i vzryvov ugol'noj pyli. Problems of early detection of fires and coal dust explosions], Problems of using Kansk-Achinsk coal at power plants: Materials. In Proceedings of the All-Russian Scientific and Practical Conference, Krasnoyarsk, Russian, 21–23 November 2000; pp. 141–149.

17.	Pavlov, A.N. Optiko-jelektronnaja sistema opredelenija trehmernyh koordinat ochaga vzryva v gazodispersnyh sistemah nachal'noj stadia [Optoelectronic system for deter-mining the three-dimensional coordinates of the explosion center in gas-dispersed systems of the initial stage]. Ph.D. Thesis, Biysk Institute of Technology, Biysk, Russian, 2010; p. 134.

18.	Pavlov, A.N.; Sypin, E.V. Optoelectronic system for determination of ignition center three-dimensional coordinates at initial stage. In Proceedings of the 9th International Conference and Seminar on Micro/Nanotechnologies and Electron Devices EDM'2010: Conference Proceedings, NSTU, Novosibirsk, Russian, 30 June–4 July 2010; pp. 417–419.

19.	Hertzberg, M.; Litton, C.D.; Donaldson, W.F.; Burgess, D. The infrared radiance and the optical detection of fires and explosions. *Symp. (Int.) Combust.* **1975**, *15*, 137–144. [CrossRef]

20.	Tupikina, N.Y.; Sypin, E.V.; Lisakov, S.A.; Pavlov, A.N.; Leonov, G.V. Two spectral retios optic electronic instrument operational parameters experimental test. *Ind. Saf.* **2015**, *4*, 66–72.

21.	Dzhigrin, A.V. Analysis of Explosion Localization Systems Operating in Coal Mines and Evaluation of the Effectiveness of Their Application. Available online: http://asvplv.ru/doc/expert_mvk.pdf (accessed on 19 January 2022).

22.	Kaledina, N.O.; Malashkina, V.A. Indicator assessment of the reliability of mine ventilation and degassing systems functioning. *J. Min. Inst.* **2021**, *250*, 553–561. [CrossRef]

23.	Vasilenko, T.A.; Kirillov, A.K.; Molchanov, A.N.; Pronskii, E.A. An NMR Study of the Ratio between Free and Sorbed Methane in the Pores of Fossil Coals. *Solid Fuel Chem.* **2018**, *52*, 361–369. [CrossRef]

24.	Magunov, A.N. Spektral'naja pirometrija objektov s neodnorodnoj temperaturoj. In *Spectral Pyrometry of Objects with Inhomogeneous Temperature*; FIZMATLIT: Moscow, Russian, 2012; p. 248.

25.	Walker, C.G. *Handbook of Spectroscopy*; Wiley-VCH GmbH: Stuttgart, Germany, 2005.

26.	Siwek, R. Experimental methods for the determination of explosion characteristics of combustible dust. In Proceedings of the 3-d International Symposium on Lose Prevention and Safety Pro-motion in the Process Industries, Basel, Switzerland, 15–19 September 1980; Volume 3.

27. Determination of Explosion Characteristics of Dust Clouds. Part 3: Determination of the Lower Explosion Limit LEL of Dust Clouds BS EN 14034-3:2006+A1:2011. Available online: https://www.en-standard.eu/bs-en-14034-3-2006-a1-2011-determination-of-explosion-characteristics-of-dust-clouds-determination-of-the-lower-explosion-limit-lel-of-dust-clouds/ (accessed on 19 January 2022).

28. Determination of Explosion Characteristics of Dust Clouds. Part 2: Determination of the Maximum Rate of Explosion Pressure rise (dP/dt)Max of Dust Clouds BS EN 14034-2:2006+A1:2011. Available online: https://www.en-standard.eu/bs-en-14034-2-2006-a1-2011-determination-of-explosion-characteristics-of-dust-clouds-determination-of-the-maximum-rate-of-explosion-pressure-rise-dp-dt-sub-m-sub-ax-of-dust-clouds/ (accessed on 19 January 2022).

29. Standard Test Method for Minimum Explosible Concentration of Combustible Dusts ASTM E1515-14. Available online: https://standards.globalspec.com/std/3857038/astm-e1515-14 (accessed on 19 January 2022).

30. Rodionov, V.A.; Pihkonen, L.V.; Zhiharev, S.Y. Dispersion of the g-type coal dust of the Vorgashorskoe field and its influence on the thermal destruction process. *Perm J. Pet. Min. Eng.* **2017**, *16*, 350–356. [CrossRef]

31. Li, Q.; Zhai, C.; Wu, H.-J.; Lin, B.-Q.; Zhu, C. Investigation on coal dust explosion characteristics using 20 L explosion sphere vessels. *J. China Coal Soc.* **2011**, *36*, 119–124.

32. Jiang, H.; Bi, M.; Huang, L.; Zhou, Y.; Gao, W. Suppression mechanism of ultrafine water mist containing phosphorus compounds in methane/coal dust explosions. *Energy* **2022**, *239*, 121987. [CrossRef]

33. Konnov, A.A.; Dyakov, I.V.; De Ruyck, J. Measurement of adiabatic burning velocity in ethane–oxygen–nitrogen and in ethane–oxygen–argon mixtures. *Exp. Therm. Fluid Sci.* **2003**, *27*, 379–384. [CrossRef]

34. Chen, C.-F.; Shu, C.-M.; Wu, H.-C.; Ho, H.-H.; Ho, S.-P. Ethylene gas explosion analysis under oxygen-enriched atmospheres in a 20-liter spherical vessel. *J. Loss Prev. Process Ind.* **2017**, *49*, 519–524. [CrossRef]

35. Wang, T.; Luo, Z.; Wen, H.; Zhang, J.; Mao, W.; Cheng, F.; Zhao, J.; Su, B.; Li, R.; Deng, J. Experimental study on the explosion and flame emission behaviors of methane-ethylene-air mixtures. *J. Loss Prev. Process Ind.* **2019**, *60*, 183–194. [CrossRef]

36. Burdjugov, S.I.; Karmanov, V.V.; Halturin, V.G. Vibrational structure of the Swann bands of the electronic spectrum of radical C2. In *Chemical Physics and Mesoscopy*; Elsevier: Amsterdam, The Netherlands, 2011.

37. Wu, J.; Yan, Z.; Ye, S.; Yang, X.; Hu, D. Micro-mechanism of carbon generation from benzene's quick reaction using monochrometer. *Acta Opt. Sin.* **2007**, *27*, 1873–1876.

38. Wang, H.; Sheen, D.A. Combustion kinetic model uncertainty quantification, propagation and minimization. *Prog. Energy Combust. Sci.* **2015**, *47*, 1–31. [CrossRef]

39. Abiev, Z.A.; Rodionov, V.A.; Paramonov, G.P.; Chernobaj, V.I. Methodology for studying the effect of inhibitory and phlegmatizing additives on the flammability and explo-siveness of coal dust. *Gorn. Inf.-Analiticheskij Bjulleteń* **2018**, *5*, 26–34. [CrossRef]

40. Snegirev, A.Y.; Tsoy, A.S. Treatment of local extinction in CFD fire modeling. *Proc. Combust. Inst.* **2015**, *35*, 2519–2526. [CrossRef]

Article

High Spatial and Temporal Resolution Bistatic Wind Lidar

Paul Wilhelm *, Michael Eggert, Julia Hornig and Stefan Oertel

Physikalisch-Technische Bundesanstalt, Department 1.4 Gas Flow, Bundesallee 100, 38116 Braunschweig, Germany; michael.eggert@ptb.de (M.E.); julia.hornig@ptb.de (J.H.); stefan.oertel@ptb.de (S.O.)
* Correspondence: paul.wilhelm@ptb.de

Abstract: The high-resolution bistatic lidar developed at the Physikalisch-Technische Bundesanstalt (PTB) aims to overcome the limitations of conventional monostatic lidar technology, which is widely used for wind velocity measurements in wind energy and meteorology applications. Due to the large measurement volume of a combined optical transmitter and receiver tilting in multiple directions, monostatic lidar generally has poor spatial and temporal resolution. It also exhibits large measurement uncertainty when operated in inhomogeneous flow; for instance, over complex terrain. In contrast, PTB's bistatic lidar uses three dedicated receivers arranged around a central transmitter, resulting in an exceptionally small measurement volume. The coherent detection and modulation schemes used allow the detection of backscattered, Doppler shifted light down to the scale of single aerosols, realising the simultaneous measurement of all three wind velocity components. This paper outlines the design details and theory of operation of PTB's bistatic lidar and provides an overview of selected comparative measurements. The results of these measurements show that the measurement uncertainty of PTB's bistatic lidar is well within the measurement uncertainty of traditional cup anemometers while being fully independent of its site and traceable to the SI units. This allows its use as a transfer standard for the calibration of other remote sensing devices. Overall, PTB's bistatic lidar shows great potential to improve the capability and accuracy of wind velocity measurements, such as for the investigation of highly dynamic flow processes upstream and in the wake of wind turbines.

Keywords: wind lidar; Doppler lidar; bistatic; metrology; traceability; wind energy; meteorology

Citation: Wilhelm, P.; Eggert, M.; Hornig, J.; Oertel, S. High Spatial and Temporal Resolution Bistatic Wind Lidar. *Appl. Sci.* **2021**, *11*, 7602. https://doi.org/10.3390/app11167602

Academic Editor: Anming Hu

Received: 16 July 2021
Accepted: 17 August 2021
Published: 19 August 2021

Publisher's Note: MDPI stays neutral with regard to jurisdictional claims in published maps and institutional affiliations.

Copyright: © 2021 by the authors. Licensee MDPI, Basel, Switzerland. This article is an open access article distributed under the terms and conditions of the Creative Commons Attribution (CC BY) license (https://creativecommons.org/licenses/by/4.0/).

1. Introduction

Accurate wind velocity measurements are an essential prerequisite for many applications in the field of wind energy and meteorology, such as wind potential analysis [1], the power curve evaluations of wind turbines [2] and atmospheric turbulence analysis [3]. For example, low measurement uncertainties are especially desired for reliable resource assessments of projected wind farms because the wind turbine power output scales with the third power to the wind velocity [4]. Wind met masts with cup and sonic anemometers are necessarily used for these applications [5]; however, they have the disadvantage of being inherently invasive and are thus prone to causing flow distortion effects [6]. Furthermore, taller masts covering hub heights of modern wind turbines are becoming economically less viable. Accordingly, ground-based wind lidar technology has become an alternative to wind met masts in the past few years, providing the distortion-free remote measurement of true wind velocity [7,8].

Conventional *monostatic* wind lidar systems measure the wind velocity component in the direction of a common transmitting and receiving beam, utilising the Doppler shift of scattered light from aerosols passing the transmitting laser beam [9]. Monostatic lidar systems provide reliable results when operated over flat terrain and in undisturbed—that is, homogeneous—flow [10]; however, these systems are not well suited for measurements over complex terrain. This is because the monostatic measurement principle can lead to measurement uncertainties in the order of 10% when operated in inhomogeneous flow [11,12]. Figure 1 illustrates this problem of the monostatic measurement principle resulting from the calculation of the horizontal velocity based on the difference of varying radial wind velocities measured at different locations and times within a single measurement cycle. The measurement beam within the measurement cycle is tilted in different directions around a circle with a height-dependent diameter of up to 100 m. Due to the dependence of the measurement uncertainty on the flow conditions at the measurement site, monostatic lidar systems are generally not traceable to the SI units. They are thus not permitted to be used as transfer standards for the calibration of other remote sensing devices [13].

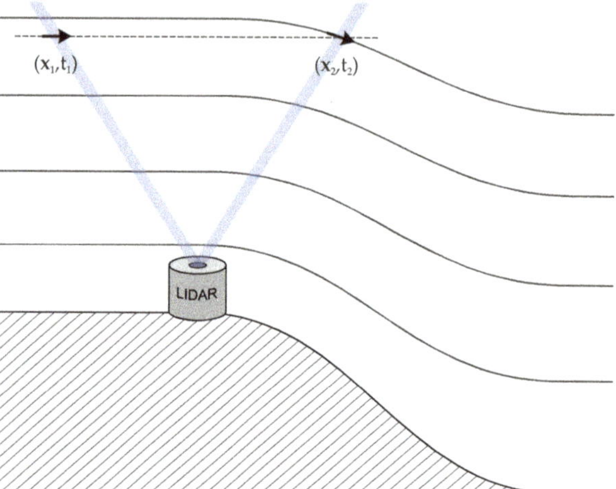

Figure 1. A monostatic lidar generally introduces a large measurement uncertainty when operating over complex terrain. This is the result of calculating the horizontal velocity based on the difference of varying radial wind velocities measured at different locations and times within a single measurement cycle in which the measurement beam is tilted in different directions.

To overcome the aforementioned limitations and provide accurate and traceable measurements over any terrain, a *bistatic* wind lidar system has been developed at PTB. Figure 2 highlights the difference in operation of both lidar technologies. The conventional monostatic lidar (cf. Figure 2a) uses a rotating prism above a single combined transmitter and receiver unit scanning over a conical contour at different locations and times, with the beam's measurement volume extending up to 30 m in length [14]. In contrast, the bistatic PTB lidar (cf. Figure 2b) uses three dedicated receivers arranged at a radius of 1 m around a central transmitter, with all units focused into a spatially highly resolved, ellipsoid-shaped measurement volume (at a height of 5 m: diameter 300 μm, length 2 mm; at a height of 200 m: diameter 12 mm, length 4 m). This facilitates the simultaneous measurement of the three-dimensional wind velocity with exceptionally high spatial resolution, down to the scale of single aerosols. In consequence, the measurement uncertainty is largely independent of flow conditions at the measurement site.

Generally, the accuracy of lidar measurements is affected by atmospheric phenomena, as the effects of absorption, refraction and dispersion become increasingly significant with long-range measurements. During conditions of haze and only light fog, the laser light is scattered from an increased number of aerosols within the measurement volume, enhancing the signal rate. Depending on the particle density, forward scattering along the optical path can cause a multitude of spurious detections, effectively lowering the measurement data rate while increasing the measurement uncertainty. Ultimately, during conditions of very heavy fog, measurement fails completely due to the intense scattering and absorption of the laser light. Contrastingly, rain affects the measurement only slightly, since raindrop signals can be easily filtered out due to the significantly different vertical velocity of the raindrops. The impact of atmospheric refraction is difficult to investigate in practice, but it was calculated to be negligible with respect to major factors influencing the total measurement uncertainty. For the signal bandwidths used with the PTB bistatic lidar, the impact of dispersion is expected to be practically negligible.

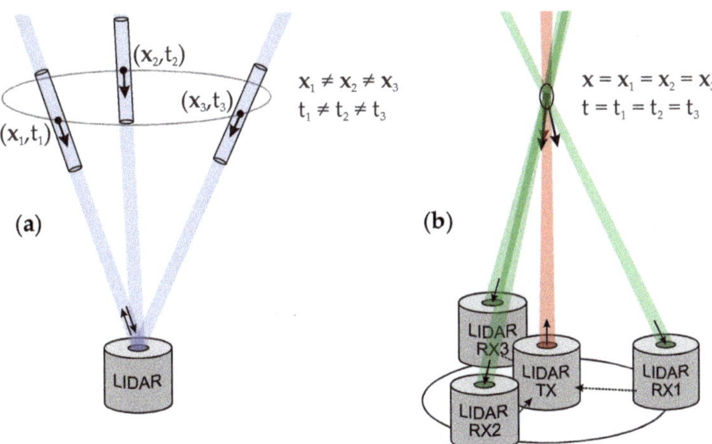

Figure 2. (**a**) A conventional monostatic lidar uses a single combined transmitter and receiver, resulting in a large measurement volume which needs to be multiplexed spatially and temporally by sequentially tilting the beam and sampling the wind speed. (**b**) The PTB bistatic lidar uses three separate receivers (*RX1*, *RX2*, *RX3*) around a central transmitter (*TX*), facilitating the simultaneous measurement of the three wind velocity components with high spatial and temporal resolution.

PTB's high-resolution bistatic lidar facilitates non-invasive wind velocity measurements at a sampling rate of up to 10 Hz (depending on the weather conditions at the measurement height) and is fully traceable to the SI units based upon the set-up's geometry, laser wavelength, and the digital time base used for frequency measurements. The detailed evaluation of the measurement uncertainty is currently in progress. Preliminary results indicate that the measurement uncertainty amounts to less than 1% at sufficiently high rates of detected particles.

2. System Description

The PTB high-resolution bistatic lidar is mounted on a custom-designed trailer, providing increased mobility and suitability for long-term outdoor use. The hinged cover is closed during field operation but may be opened during development and servicing, as shown in Figure 3. The sensitive optical set-up and the signal processing unit are located under the cover. To prevent vibrations during transport and operation, the optical set-up is mechanically decoupled from the trailer using air suspension. Further, to prevent the mechanical expansion and torsion of the optics carrier aluminium profile, the entire set-up is climatised to about 20 °C using a 2 kW water-cooling unit. The transmitting laser amplifier is encased separately and water-cooled to reduce the excessive heating of the remaining set-up. Additional fans ensure sufficient air convection, and a dehumidifier prevents condensation forming on the optical windows inside the cover.

The optical receivers are located at a radius of 1 m around the central transmitter in order to provide both sufficient backscattered light intensity (quasi-backwards; i.e., in a quasi-vertical direction) and sufficient resolution for determining the horizontal wind velocity component. Each receiver includes a servo-driven focus lens and a tilting mirror, as shown in Figure 4. The tilting mirror is driven by a servo-piezo ensemble in order to provide sufficient resolution for the accurate positioning of the Gaussian beams at measurement heights of up to 250 m. As the transmitter does not need to be tilted, it only consists of a focus lens and a fixed mirror. Optically adjusting and focusing all the receivers and the transmitter into the small measurement volume is achieved by means of a specially developed scanning algorithm, successively approximating motor positions based on the received light intensity and measurement height.

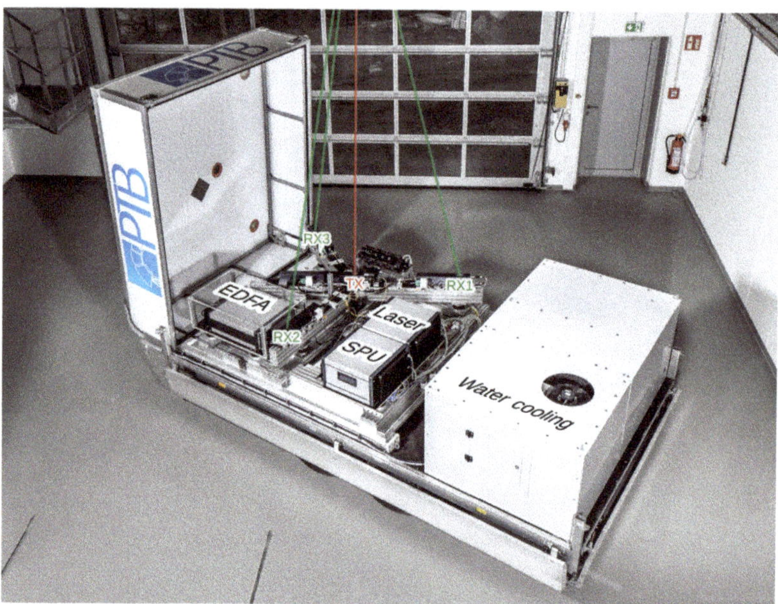

Figure 3. The PTB high-resolution bistatic lidar is mounted on a custom-designed trailer with a hinged cover (opened here), under which the optical set-up and the signal processing unit (SPU) are located.

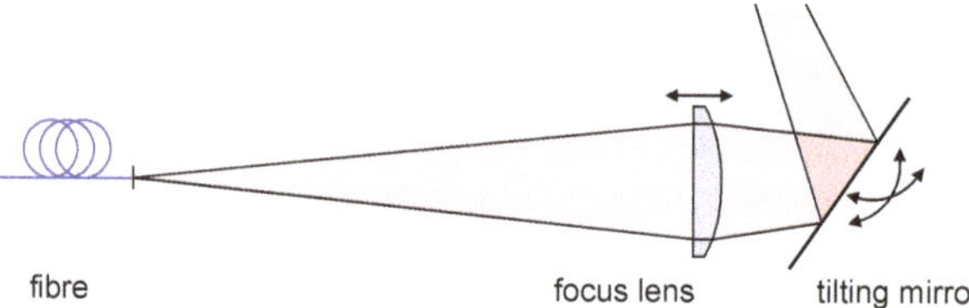

Figure 4. Principle of the optomechanical set-up: Each receiver consists of a servo-driven focus lens and a tilting mirror which is driven by a servo-piezo ensemble in order to provide sufficient resolution for accurate positioning at measurement heights up to 250 m. As the transmitter does not need to be tilted, it only consists of a focus lens and a fixed mirror.

The PTB high-resolution bistatic lidar uses coherent detection, processing the interference of the transmitted and backscattered light of aerosols carried along the flow [15]. While bistatic lidar technology generally poses the problem of low backscattered signal intensity, the received Doppler spectrum is concentrated into a significantly narrower bandwidth compared to that of conventional monostatic lidar. This is due to the more uniform motion of aerosols within the small measurement volume, allowing the detection of Doppler peaks with an improved signal-to-noise ratio (SNR) [16]. At the receiver, the backscattered light from multiple aerosols with slightly different Doppler frequencies is detected [17]. Correlation techniques are used during signal processing, ensuring that only Doppler frequencies emitted by the same aerosol are evaluated [18]. The measurement height is first roughly determined by the theoretical height, which is set using the optomechanical actuators, and it is finely determined by means of the difference in coherent phases of both Doppler shifted modulation peaks; due to the used modulation scheme, the phases are periodic over the beam's length.

While the vertical wind velocity component directly generates an absolute Doppler shift to the backscattered light, the horizontal wind velocity component must be computed by means of correlated frequency offsets between all three receiving channels. Overall, due to the sharp angle between the transmitter beam and the receiver beam, the detection of the vertical wind velocity component is orders of magnitude more sensitive than that of the horizontal wind velocity component [18]. With an average beam diameter of 10 mm between heights of 100 m to 200 m, the average transit time of a particle passing the measurement volume at a speed of $10 \ \mathrm{m \cdot s^{-1}}$ amounts to 1 ms. Both the reference laser's maximum bandwidth and the frequency resolution of the signal processing were chosen accordingly.

The optical set-up is constructed using single-mode fibre optics internally, as shown in Figure 5. A narrow-bandwidth (<1 kHz) laser reference of a wavelength of 1550 nm is modulated using an acousto-optic modulator (AOM), amplified (up to 30 W CW) using an erbium-doped fibre amplifier (EDFA), and coupled into the transmitter beam optics. The received light is fed into fibre couplers, where it is coherently mixed (heterodyned) with the light of the reference laser; the resulting beat frequencies are the Doppler frequencies, converted down into a lower frequency range. Each optical mixing product is fed into a balanced photodetector (PD), where it is converted to an electrical signal that is ready to be processed.

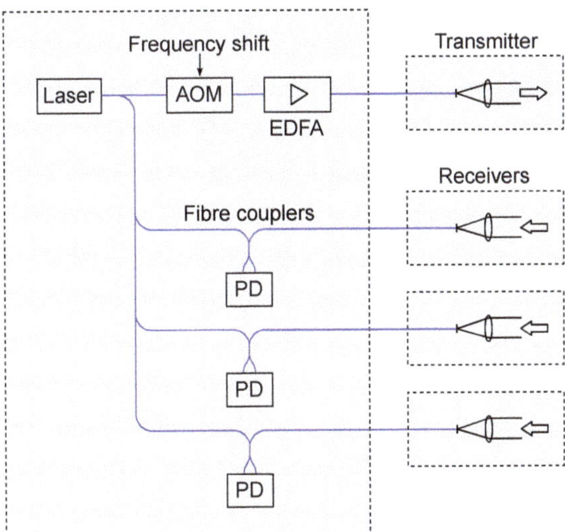

Figure 5. The optical set-up is constructed using fibre optics. A narrow-bandwidth laser reference is modulated, amplified and fed into the transmitter beam optics. The light scattered back into the receiving beam optics is fed into fibre couplers, where it is coherently mixed with the reference laser's light. Each optical mixing product is fed into a balanced photodetector, where it is converted to an electrical signal.

The signal processing is performed in two steps: a field-programmable gate array (FPGA) pre-processes the signals at a high data rate, and a CPU post-processes these signals at a lower data rate. Both units are contained inside a common MicroTCA system carrier, transferring data via a high-speed backplane protocol. The FPGA has four analogue-digital converters (ADC) and four digital-analogue converters (DAC) associated with it, all interfaced using a fixed-latency protocol to ensure deterministic phase relationships within the coherent signal chain. The AOM modulation signal is provided using one of the available DACs, and two spectral peaks at 75 MHz ± 12.5 MHz are added to the transmitted light intensity. These are necessary for determining the Doppler polarity, signal phase and, accordingly, the measurement height by means of spectral signal processing. Three of the ADCs are used for digitising the received signals, as shown in Figure 6. The electrical signals from each balanced photodetector are subtracted and digitised using one dedicated ADC sampling at 300 MHz. Using the FPGA, the signal is mixed with a complex valued 75 MHz reference signal. The signal is then low-pass filtered and subsampled at a rate of 50 MHz. Finally, the signal is transmitted to the CPU, where it is transformed into the spectral domain using a fast Fourier transform (FFT) with a block length of 32,768 samples, taking about 1 ms to complete. Further signal processing, such as filtering and correlating, typically increases the total computing time per sampling block to about 7 ms (which is only about 10% of the theoretically achievable real-time performance).

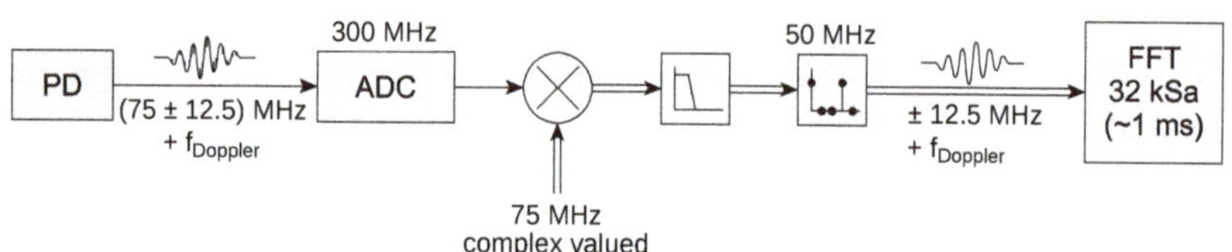

Figure 6. The electrical signals from each balanced photodetector are digitised using one dedicated ADC sampling at 300 MHz. The signal is then mixed with a complex valued 75 MHz reference signal, low-pass filtered and subsampled at a rate of 50 MHz. Finally, the signal is transformed into the spectral domain using an FFT with a block length of 32,768 samples.

3. Validation and Characterisation

Since the development of PTB's high-resolution bistatic lidar started in 2010, several comparative measurements have been conducted in order to characterise this remote sensing device over a range of different operating conditions. Some of these measurements are described in the following sections.

3.1. PTB Bistatic Lidar/Ultrasonic Anemometer (Vaisala WMT700)

In 2014, the first comparative measurements between the PTB bistatic lidar and a Vaisala WMT700 ultrasonic anemometer mounted at a height of 10 m were conducted [19]. Being representative of complex terrain, a common measurement site amidst several buildings at PTB was chosen.

Considering the disturbed flow conditions and the distance of about 50 cm between the ultrasonic anemometer and the PTB bistatic lidar's measurement volume, the comparative measurements indicated a very good agreement for both wind speed and wind direction. Orthogonal linear regressions for the 1 min averages of wind speed and wind direction yielded Pearson correlation coefficients of 0.982 and 0.905, respectively.

3.2. PTB Bistatic Lidar/Monostatic Lidar (WindCube)

In 2014, two comparative measurements between the PTB bistatic lidar and a monostatic lidar (WindCube) were conducted over flat terrain at PTB's antenna testing ground, and over complex terrain amidst several buildings at PTB [20]. At each location, both devices were positioned at a distance of 10 m from each other and set to a measurement height of 100 m. Over flat terrain, both devices showed good agreement with deviations below 1% over a one-hour period of low atmospheric turbulence. However, over complex terrain, both devices showed significant deviations in the order of 15%, as it could be expected for measurements at different locations in turbulent flow and compared against conventional monostatic lidar technology.

3.3. PTB Bistatic Lidar/Cup Anemometers (135 m Wind Met Mast)/Monostatic Lidar (WindCube)

In 2015 and 2016, two comparative measurements between the PTB bistatic lidar and a 135 m met mast equipped with several cup anemometers were conducted at the

Deutsche WindGuard testing ground in Aurich, Germany [21,22]. In 2015, a cup anemometer mounted on a boom at a height of 100 m was used for reference, and the PTB bistatic lidar was positioned nearby at a distance of about 1 m. In this situation, both devices were exposed to the *disturbed wake flow* of a nearby wind turbine. In 2016, the cup anemometer at the top of the met mast (135 m) was used for reference and the PTB bistatic lidar was positioned next to the met mast at a distance of about 3 m. For this second campaign, monostatic lidar (WindCube) data were also available and were incorporated into the comparative measurements. Wind velocity data were evaluated according to IEC 61400-12-1 [5], sorting horizontal wind speeds into discrete bins with a resolution of $0.5\,\mathrm{m \cdot s^{-1}}$. Generally, large error bars indicate that few measurement samples were available in the corresponding wind speed ranges.

Figure 7 shows the deviation between the PTB bistatic lidar and the boom-mounted anemometer in *disturbed flow* and for 1 s averaging intervals. Here, the PTB bistatic lidar shows deviations below 1% over a large range of wind speeds from $6\,\mathrm{m \cdot s^{-1}}$ to $12\,\mathrm{m \cdot s^{-1}}$ when averaged over 1 s intervals. Figures 8 and 9 show the deviations between the PTB bistatic lidar and the top anemometer and between the monostatic lidar (WindCube) and the top anemometer in *undisturbed flow* for 10 min and 1 min averaging intervals, respectively. In *undisturbed flow*, for 10 min averaging intervals and within bins in which at least 10 wind speeds were sampled, the PTB bistatic lidar's deviation with respect to the top anemometer reference amounts to less than 0.5% (cf. Figure 8), which is well within the cup anemometer's measurement uncertainty (roughly between $\pm 1\%$ at $4\,\mathrm{m \cdot s^{-1}}$ and $\pm 0.7\%$ at $13\,\mathrm{m \cdot s^{-1}}$). Under the same conditions, the monostatic lidar shows deviations up to -1.5% (cf. Figure 8). In *undisturbed flow* and for 1 s averaging intervals, the monostatic lidar shows large deviations (cf. Figure 9). This is because it resorts to interpolation for intervals shorter than its native sampling interval, which comprises multiple seconds for a single scan. Contrastingly, under the *same conditions*, the PTB bistatic lidar's deviation only increases significantly beyond 1% at very low or very high wind speeds. Extensive investigations and mathematical simulations showed that this is due to the different dynamic responses of the used sensing devices employed, and also due to the limited correlation of wind velocities sampled at slightly different locations [21].

Overall, owing to its high spatial and temporal resolution, the PTB bistatic lidar has been proven to achieve lower measurement uncertainty than conventional lidar systems, independent of the averaging time.

3.4. PTB Bistatic Lidar/LDA in Wind Tunnel

In 2018, the PTB bistatic lidar was characterised in the wind tunnel at PTB's Competence Center for Wind Energy (CCW). Being specifically constructed for the validation of the PTB bistatic lidar, the wind tunnel is mounted on a platform at a height of 8 m, allowing the PTB bistatic lidar trailer to be positioned immediately below it to sense its velocity field.

It was shown that the wind tunnel provides a well-defined, homogeneous velocity field suitable for calibrations. The wind tunnel shows a turbulence level of less than 0.35% for flow speeds ranging from $2\,\mathrm{m \cdot s^{-1}}$ to $30\,\mathrm{m \cdot s^{-1}}$. At distances from the nozzle below 375 mm, where the measurements take place, it shows deviations of the flow velocity below $0.15\% \cdot \mathrm{dm^{-1}}$ [23].

A laser Doppler anemometer (LDA) was used as a reference standard, as LDA devices have measurement uncertainties below 0.2% [13], which is lower than the measurement uncertainties achievable with cup anemometers (roughly between $\pm 0.7\%$ and $\pm 1\%$). Figure 10 shows the deviation of the PTB bistatic lidar from the LDA. These first validation measurements of wind speeds ranging from $4\,\mathrm{m \cdot s^{-1}}$ to $16\,\mathrm{m \cdot s^{-1}}$ showed an average deviation of 0.37% between the LDA and the PTB high-resolution bistatic lidar [24,25].

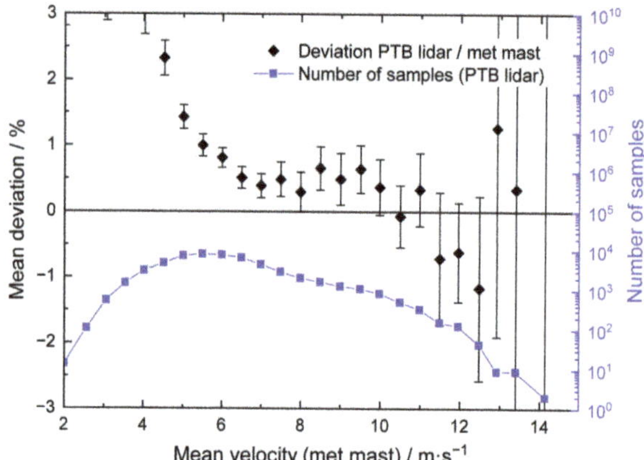

Figure 7. Results of the 2016 comparative measurement between the PTB bistatic lidar and a cup anemometer mounted on a met mast boom at a height of 100 m in *disturbed flow* at a measurement height of 100 m, evaluated according to IEC 61400-12-1 using 1 s intervals. The PTB bistatic lidar exhibits absolute deviations below 1% over a large range of wind speeds from $6 \text{ m} \cdot \text{s}^{-1}$ to $12 \text{ m} \cdot \text{s}^{-1}$. The PTB bistatic lidar's deviation only increases significantly beyond 1% at very low or very high wind speeds. Extensive investigations and mathematical simulations showed that this is due to the different dynamic responses of the sensing devices employed, and also due to the limited correlation of wind velocities sampled at slightly different locations.

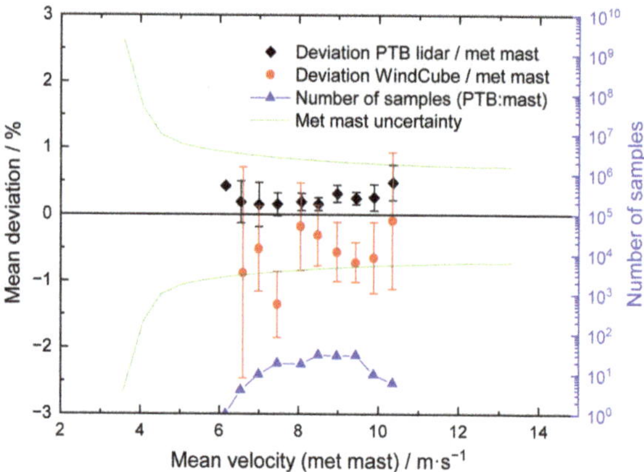

Figure 8. Results of the 2016 comparative measurement between the PTB bistatic lidar and the top anemometer and between the monostatic lidar (WindCube) and the top anemometer in *undisturbed flow* at a measurement height of 135 m, evaluated according to IEC 61400-12-1 using 10 min intervals. For bins in which at least 10 values were obtained, the PTB bistatic lidar's deviation is less than 0.5%, which is well within the cup anemometer's measurement uncertainty (roughly between ±1% at $4 \text{ m} \cdot \text{s}^{-1}$ and ±0.7% at $13 \text{ m} \cdot \text{s}^{-1}$). Under the same conditions, the monostatic lidar shows deviations up to -1.5%.

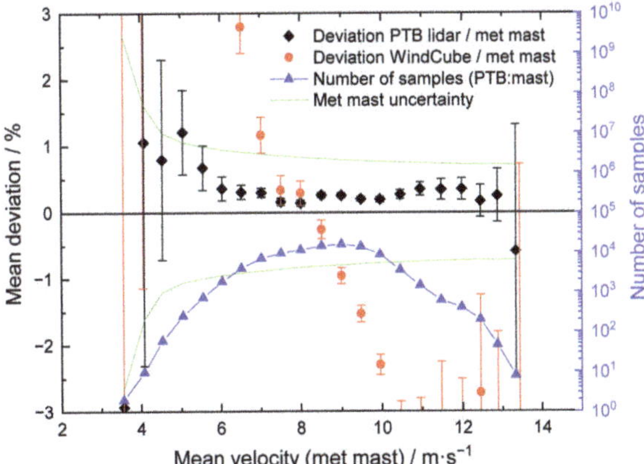

Figure 9. Results of the 2016 comparative measurement between the PTB bistatic lidar and the top anemometer and between the monostatic lidar (WindCube) and the top anemometer in *undisturbed flow* at a measurement height of 135 m, evaluated according to IEC 61400-12-1 using 1 s intervals. The monostatic lidar shows large deviations, while the PTB bistatic lidar's deviation only increases significantly beyond 1% at very low or very high wind speeds.

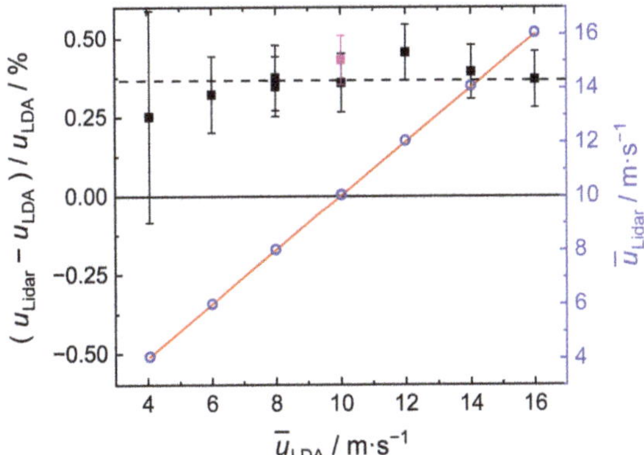

Figure 10. Results of the 2018 comparative measurement between the PTB bistatic lidar and an LDA in the wind tunnel at PTB's Competence Center for Wind Energy, showing an average deviation of 0.37% between both devices. In order to verify that the PTB bistatic lidar is free of any angular dependence, an additional sample was recorded at $10\ \mathrm{m \cdot s^{-1}}$ (shown in pink) after the PTB bistatic lidar trailer was rotated 90°.

3.5. PTB Bistatic Lidar/Sonic Anemometer (CSAT3B) in Turbulent Flow

In 2019, a comparative measurement between the PTB bistatic lidar and a CSAT3B ultrasonic anemometer mounted on top of a 30 m mast was conducted over flat terrain at the site of the Johann Heinrich von Thünen Institut (the German Federal Research Institute for Rural Areas, Forestry and Fisheries) in Braunschweig, Germany [3]. At this height, the PTB bistatic lidar's measurement volume (diameter $d = 2\ \mathrm{mm}$ and length $l = 50\ \mathrm{mm}$) is comparable to that of the used sonic anemometer. The PTB bistatic lidar was positioned 9 m away from the mast, and both devices sampled wind velocity data at a rate of 10 Hz over a period of about two weeks.

Figure 11 shows the orthogonal linear regressions of the average horizontal wind velocity component $\overline{u}$, of the fluctuation of the vertical wind velocity component $\overline{w'w'}^{1/2}$, and of the shear stress velocity u_*. Table 1 shows the corresponding statistical parameters.

Both devices showed very good agreement, especially for the fluctuation of the vertical wind velocity component, with a mean deviation of only 0.017 m · s^{-1} at an intercept of −0.009 m · s^{-1} and a slope of 0.989. There was also good agreement for the average horizontal wind velocity component $\overline{u}$ with an intercept of 0.044 m · s^{-1} and a slope of also 0.989. However, at a slope of 0.973, the shear stress velocity u_* exhibited a slightly larger deviation from unity. This is demonstrably due to the angular dependence inherent to ultrasonic anemometers.

The dynamics of fully developed turbulence in the inertial subrange (an intermediate range of scales within the underlying energy cascade) can be described by certain similarity laws [26]. Specifically, the ensemble cospectrum (spectrum of the cross-correlation of two orthogonal velocity components) follows a −7/3 power law [27]. Figure 12 shows the ensemble cospectra Co_{uw} between the wind velocity components u and w of the PTB bistatic lidar and the CSAT3B sonic anemometer. The PTB high-resolution bistatic lidar is in excellent agreement with the theoretical distribution, showing a drop-off in cospectral power density consistent with the theoretical −7/3 law. However, due to its angular dependence, at higher frequencies, the CSAT3B sonic anemometer deviates significantly from the theoretical distribution.

Overall, although both the PTB bistatic lidar and the employed ultrasonic anemometer have proven to be well suited for measurements in turbulent flow, the PTB bistatic lidar is the more favourable option if high precision is desired. This is due to its distortion-free measurement capability, which is free of any angular dependence.

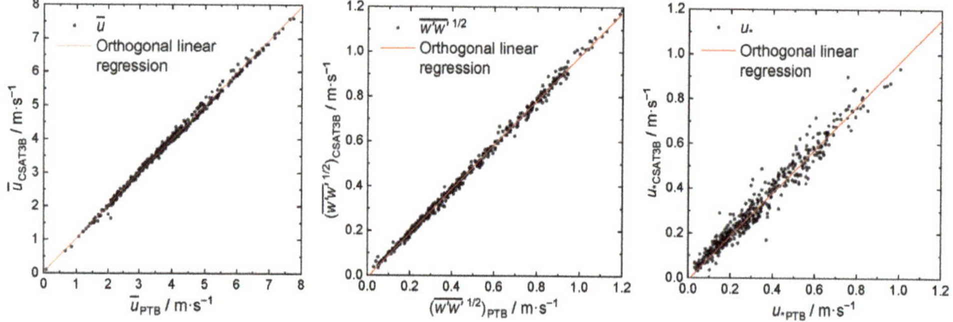

Figure 11. Results of the 2019 comparative measurement between the PTB bistatic lidar and a CSAT3B ultrasonic anemometer mounted on top of a 30 m mast in turbulent flow, showing orthogonal linear regressions of the average horizontal wind velocity component $\overline{u}$, of the fluctuation of the vertical wind velocity component $\overline{w'w'}^{1/2}$ and of the shear stress velocity u_*.

Table 1. Statistical parameters of the 2019 comparative measurement between the PTB bistatic lidar and a CSAT3B ultrasonic anemometer mounted on top of a 30 m mast.

	$\overline{u}$	$\overline{w'w'}^{1/2}$	u_*
Bias (m · s^{-1})	0.003	−0.009	−0.009
Mean deviation, RMSE (m · s^{-1})	0.082	0.017	0.042
Regression: intercept (m · s^{-1})	0.044	−0.009	0.000
Regression: slope	0.989	0.989	0.973
Pearson correlation coefficient	0.998	0.998	0.980

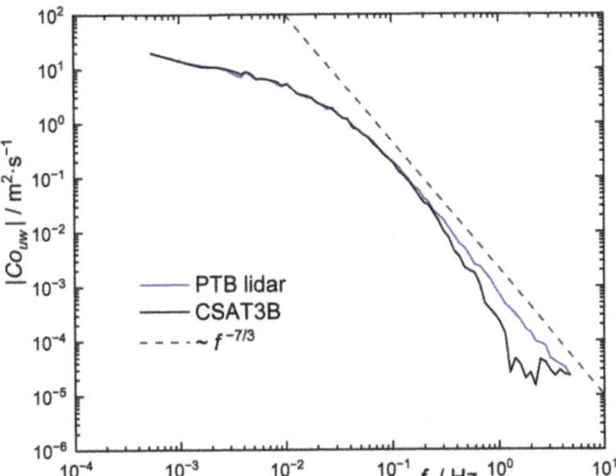

Figure 12. Results of the 2019 comparative measurement between the PTB bistatic lidar and a CSAT3B ultrasonic anemometer mounted on top of a 30 m mast in turbulent flow, showing the ensemble cospectra Co_{uw} between the wind velocity components u and w (absolute values) of the PTB bistatic lidar and the CSAT3B sonic anemometer. The dashed line indicates the theoretical $-7/3$ power law in the inertial subrange.

3.6. PTB Bistatic Lidar/Cup Anemometer (200 m Wind Met Mast)

In 2020, a comparative measurement between the PTB bistatic lidar and a Thies Clima First Class Advanced cup anemometer mounted on top of a 200 m met mast was conducted near the summit of the Rödeser Berg near Kassel, Germany. Both the PTB bistatic lidar and the top cup anemometer were measuring at a height of 200 m for about three weeks with air temperatures ranging from -2.1 °C to 21.1 °C at a height of 10 m and from -2.7 °C to 17.8 °C at a height of 187 m. Additionally, wind direction data were obtained from a Thies Clima First Class wind vane mounted on a boom of the wind met mast at a height of 187 m. In order to enable a reasonable comparison, this particular measurement campaign required the use of filtering operations, removing periods of fog during which the lidar measurements were significantly impaired. Furthermore, to enable an evaluation according to IEC 61400-12-1, effects of turbulence due to the wakes of the nearby wind turbines were filtered out. Because of the low detection rate of particles at the measurement height of 200 m, 10 min averages are used throughout the following evaluation.

Figure 13 shows the overall measurement site, which includes two wind turbines (a third wind turbine located south-east is not shown). The PTB bistatic lidar trailer was positioned next to the met mast at a distance of 4.6 m. All angles stated in this section are referenced to the 0° north direction, counting in a clockwise direction (meteorological wind direction). Multiple pairs of oppositely mounted booms are located at different heights on the wind met mast at 145° and 325° angles. The wind turbines are located at 146° and 307° angles.

Figure 14 shows the deviation of the PTB bistatic lidar to the top anemometer for *unfiltered* 10 min averaged wind speeds larger than 4 m·s⁻¹. However, these unfiltered samples are problematic in two ways. Firstly, large perturbation effects due to the wakes of the nearby wind turbines are visible around 146° and 307° angles, and secondly, due to the PTB bistatic lidar's optical measurement principle, fog occasionally causes large, non-systematic deviations which are clearly visible between 90° and 270° angles. Thus, to enable a more reasonable comparison, the data were filtered around the angles corresponding to the wakes of the nearby wind turbines, and periods of fog were filtered using an upper limit on the backscattered signal amplitude. Figure 15 shows the deviation between the PTB bistatic lidar and the top anemometer for *filtered* 10 min averages, indicating good agreement between both instruments with absolute deviations below 2.5%.

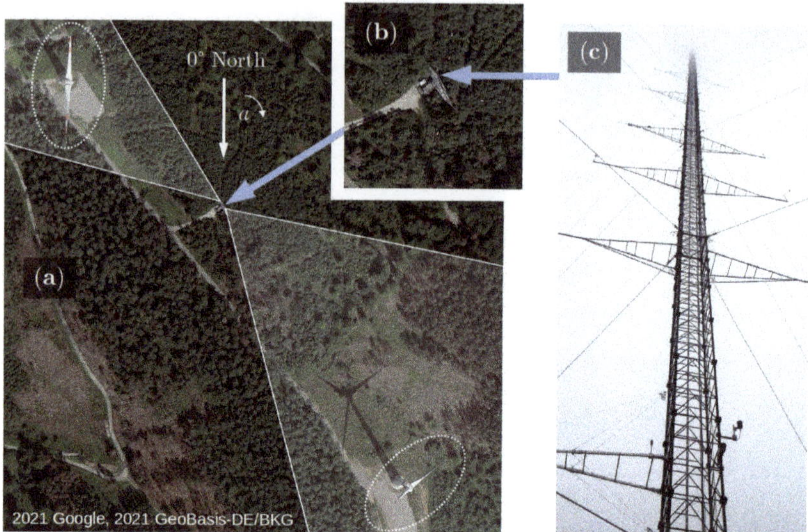

Figure 13. (**a**) Measurement site including two wind energy turbines located at 146° and 307° angles (a third wind turbine located south-east is not shown). Around the met mast, the areas corresponding to the wakes of the nearby wind turbines are highlighted between 100° to 165° and 285° to 330° angles. (**b**) Slightly larger view of the wind met mast and its orientation with respect to the measurement site. (**c**) Met mast with multiple booms mounted at different heights and at 145° and 307° angles.

Figure 16 shows the deviation between the PTB bistatic lidar and the top anemometer for filtered wind speeds ranging from 4 m·s^{-1} to 19.5 m·s^{-1}. Here, wind velocity data were evaluated according to IEC 61400-12-1 [5], sorting horizontal wind speeds into discrete bins with a resolution of 0.5 m·s^{-1}. Large error bars indicate that few measurement samples were available in the corresponding wind speed ranges. Again, both the PTB bistatic lidar and the top anemometer show good agreement, with mean deviations below 0.5% over a large range of wind speeds. Figure 17 shows the linear orthogonal regression of the wind directions measured by the PTB bistatic lidar and the wind vane mounted on the met mast. With an intercept of 3.736° and a slope of 0.996, both instruments are in excellent agreement regarding the measured wind direction.

Generally, a directional dependence of wind velocity measurements with respect to met mast booms has been extensively covered in the literature and is known as the *mast and boom effect*. Firstly, this effect causes large deviations in the wind velocity within a narrow angular range, when the anemometer is located in the lee of the mast. Secondly, an additional and more or less sinusoidal deviation in the range of ±1% to ±2% (for lattice towers) is induced over the entire angular range [6,28,29]. In order to mitigate the mast and boom effect specifically, this comparative measurement was conducted with respect to the top anemometer located at 200 m, even though there were many periods of fog at this measurement height.

Overall, the directional effects due to the wakes of the nearby wind turbines and the PTB bistatic lidar's sensitivity to periods of fog necessitated the additional filtering of the data obtained in this particular comparative measurement campaign. Nevertheless, both the PTB bistatic lidar and the wind met mast's top anemometer show good agreement with mean deviations below 0.5% over a large range of wind speeds. Furthermore, the wind directions as measured by the PTB bistatic lidar and the wind vane mounted on the met mast are in excellent agreement.

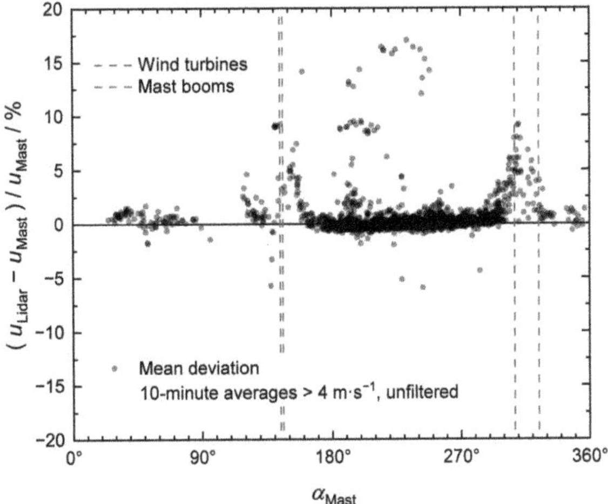

Figure 14. Results of the 2020 comparative measurement between the PTB bistatic lidar and a cup anemometer mounted on top of a 200 m met mast, showing the directional dependence of the mean deviation between the PTB bistatic lidar and the top anemometer for *unfiltered* 10 min averages. Large perturbation effects due to the wakes of the nearby wind turbines are visible around angles 146° and 307°. Additionally, fog occasionally causes very large, non-systematic deviations which are clearly visible between 90° and 270° angles.

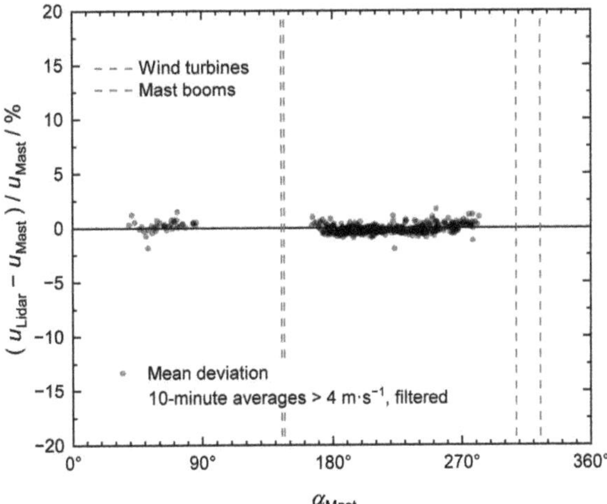

Figure 15. Results of the 2020 comparative measurement between the PTB bistatic lidar and a cup anemometer mounted on top of a 200 m met mast, showing the directional dependence of mean deviation between the PTB bistatic lidar and the top anemometer for *filtered* 10 min averages. Perturbation effects due to the wakes of the nearby wind turbines have been reduced by directional filtering. Furthermore, periods of fog were excluded from the data.

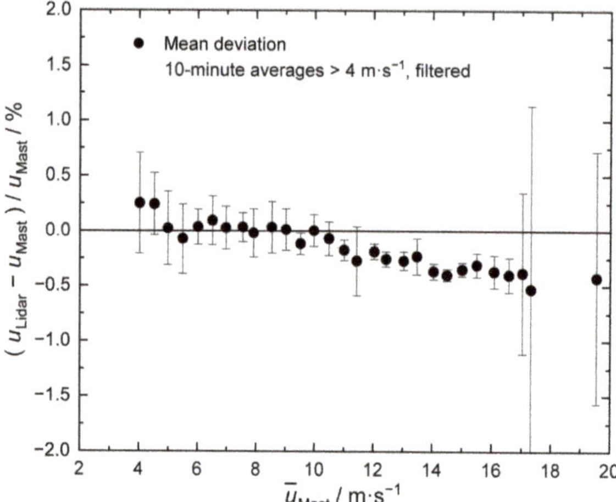

Figure 16. Results of the 2020 comparative measurement between the PTB bistatic lidar and a cup anemometer mounted on top of a 200 m met mast, evaluated according to IEC 61400-12-1 using filtered 10 min averages. Both instruments show good agreement with mean deviations between +0.25% and −0.5% over a large range of wind speeds.

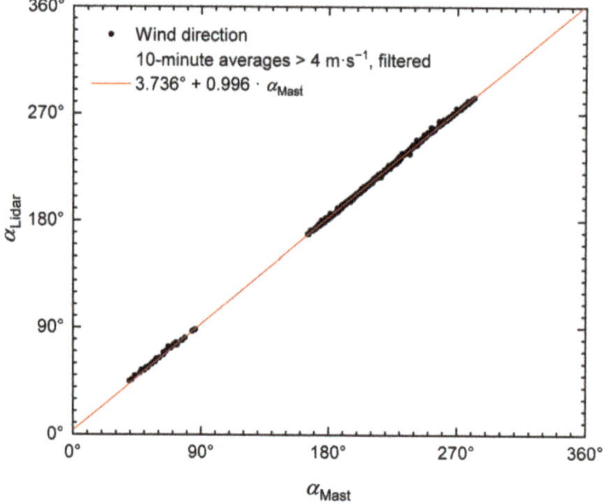

Figure 17. Results of the 2020 comparative measurement between the PTB bistatic lidar and a cup anemometer mounted on top of a 200 m met mast, showing the linear orthogonal regression of the filtered wind directions measured by the PTB bistatic lidar and the wind vane mounted on the met mast.

4. Conclusions and Outlook

This paper has outlined the design details and theory of operation of PTB's high-resolution bistatic lidar. It has also provided an overview of selected comparative measurements. The comparative measurements between the PTB bistatic lidar and the cup anemometers mounted on the met masts show that the measurement uncertainty of PTB's bistatic lidar is well within the measurement uncertainty of traditional cup anemometers (roughly between $\pm 1\%$ at 4 m $\cdot$ s^{-1} and $\pm 0.7\%$ at 13 m $\cdot$ s^{-1}) while being fully independent of its site. The comparative measurement between the PTB bistatic lidar and an industry-standard ultrasonic anemometer (CSAT3B) in turbulent flow shows that the PTB bistatic lidar provides distortion-free measurement capability which is free of any angular dependence and is in excellent agreement with the theoretical $-7/3$ power law in the inertial subrange.

To summarise, the high-resolution bistatic lidar developed at PTB has been proven to provide spatially and temporally highly resolved remote measurements of the wind velocity at heights ranging from 5 m to 200 m over any terrain. Furthermore, it provides measurement results with exceptionally low measurement uncertainty that are traceable to the SI units. Thus, it is planned to use the PTB high-resolution bistatic lidar as a transfer standard for other wind velocity remote sensing devices in the future. A full measurement uncertainty budget is currently in preparation and it is planned to be addressed in a separate paper. Overall, the PTB bistatic lidar shows great potential to enhance the capability and accuracy of measurements in applications such as wind potential analysis, the power curve evaluation of wind turbines and atmospheric turbulence analysis.

Nevertheless, there is still room for improvement. Adjusting the measurement height is currently slow, requiring multiple scans over an interval of several minutes. Therefore, the optomechanical section is being completely reworked. This is moreover being combined with an improved signal processing unit using the latest radio-frequency system-on-chip (RFSoC) technology. In the future, this will enable the fast and autonomous selection of measurement heights, allowing the PTB bistatic lidar to perform wind profile measurements. The improved signal processing will take the performance closer to real-time, increasing both the spectral resolution and the number of aerosols detectable per unit of time, which is also a requirement for fast beam alignment.

Several desirable functionalities and improvements are still missing from the PTB bistatic lidar at its current stage. Firstly, because of the PTB bistatic lidar's optical measurement principle, its measurement capability is degraded during weather conditions such as precipitation and fog. It is thus desirable to incorporate the automatic detection and mitigation of these conditions based upon the SNR and other spectral characteristics of the received Doppler spectra. Secondly, a thorough investigation of the measurement uncertainty as a function of the SNR and the characteristic distributions of correlated Doppler peaks for varying meteorological conditions is needed in order to improve upon the achieved measurement uncertainty by fine-tuning the digital signal processing chain. These are only some of the aspects that are to be addressed over the course of the coming years in order to provide a fully autonomous remote sensing device that is suitable for use as a transfer standard.

Author Contributions: Conceptualization, M.E.; methodology, M.E.; software, P.W., M.E.; validation, M.E.; formal analysis, M.E.; investigation, M.E., P.W.; data curation, M.E., P.W.; writing—original draft preparation, P.W.; writing—review and editing, P.W., M.E., J.H.; visualization, M.E., P.W., S.O.; supervision, J.H.; project administration, J.H.; funding acquisition, J.H., P.W. All authors have read and agreed to the published version of the manuscript.

Funding: This research was funded by the German Federal Ministry for Economic Affairs and Energy (BMWi), funding codes 0325416B (*WindLidarDSP*) and 0325945 (*PTB Wind*).

Data Availability Statement: The data presented in this study are available upon request from the corresponding author. The data are not publicly available due to third party agreements. Data were obtained from Deutsche WindGuard GmbH, the Karlsruhe Institute of Technology, the Carl von Ossietzky University Oldenburg and the Fraunhofer Institute for Energy Economics and Energy System Technology.

Acknowledgments: We thank Deutsche WindGuard GmbH (Varel, Germany), the Karlsruhe Institute of Technology (Garmisch-Partenkirchen, Germany), the Carl von Ossietzky University Oldenburg (Oldenburg, Germany) and the Fraunhofer Institute for Energy Economics and Energy System Technology (Kassel, Germany) for making the joint measurement campaigns possible.

Conflicts of Interest: The authors declare no conflict of interest. The funders had no role in the design of the study; in the collection, analyses, or interpretation of data; in the writing of the manuscript or in the decision to publish the results.

Abbreviations

The following abbreviations are used in this manuscript:

ADC	Analogue–digital converter
AOM	Acousto-optic modulator
CCW	Competence Center for Wind Energy
DAC	Digital–analogue converter
EDFA	Erbium-doped fibre amplifier
FFT	Fast Fourier transform
FPGA	Field-programmable gate array
LDA	Laser Doppler anemometer
Lidar	Light detection and ranging
RMSE	Root mean square error
SNR	Signal-to-noise ratio
PD	Photodetector
PTB	Physikalisch-Technische Bundesanstalt

References

1. MEASNET. Evaluation of Site-Specific Wind Conditions, Version 2. April 2016. Available online: https://www.measnet.com/wp-content/uploads/2016/05/Measnet_SiteAssessment_V2.0.pdf (accessed on 4 March 2021).
2. Scheurich, F.; Enevoldsen, P.B.; Paulsen, H.N.; Dickow, K.K.; Fiedel, M.; Loeven, A.; Antoniou, I. Improving the Accuracy of Wind Turbine Power Curve Validation by the Rotor Equivalent Wind Speed Concept. *J. Phys. Conf. Ser.* **2008**, *753*, 072029. [CrossRef]
3. Mauder, M.; Eggert, M.; Gutsmuths, C.; Oertel, S.; Wilhelm, P.; Voelksch, I.; Wanner, L.; Tambke, J.; Bogoev, I. Comparison of turbulence measurements by a CSAT3B sonic anemometer and a high-resolution bistatic Doppler lidar. *Atmos. Meas. Tech.* **2020**, *13*, 969–983. [CrossRef]
4. Kalmikov, A. Wind Power Fundamentals. Available online: http://web.mit.edu/wepa/WindPowerFundamentals.A.Kalmikov.2017.pdf (accessed on 4 March 2021).
5. IEC 61400-12-1. *Wind Energy Generation Systems—Part 12-1: Power Performance Measurements of Electricity Producing Wind Turbines*, 2nd ed.; International Electrotechnical Commission: Geneve, Switzerland, 2017.
6. Lindelöw, P.J.P.; Friis Pedersen, T.; Gottschall, J.; Vesth, A.; Wagner, R.; Schmidt Paulsen, U.; Courtney, M. *Flow Distortion on Boom Mounted Cup Anemometers*; Technical University of Denmark: Roskilde, Denmark, 2010; Volume 1738(EN).
7. Albers, A.; Janssen, A.W.; Mander, J. How to Gain Acceptance for Lidar Measurements. Available online: https://www.windguard.de/veroeffentlichungen.html (accessed on 4 March 2021).
8. Slinger, C.; Harris, M. Introduction to Continuous-Wave Doppler Lidar. Available online: http://breeze.colorado.edu/ftp/RSWE/Chris_Slinger.pdf (accessed on 4 March 2021).
9. Drain, L.E. *The Laser Doppler Technique*, 1st ed.; John Wiley & Sons: New York, NY, USA, 1980.
10. Gottschall, J.; Courtney, M.S.; Wagner, R.; Jørgensen, H.E.; Antoniou, I. Lidar profilers in the context of wind energy—A verification procedure for traceable measurements. *Wind Energy* **2012**, *15*, 147–159. [CrossRef]
11. Bradley, S. Wind speed errors for LIDARs and SODARs in complex terrain. *IOP Conf. Ser. Earth Environ. Sci.* **2008**, *1*, 012061 [CrossRef]
12. Bingöl, F.; Mann, J.; Foussekis, D. Conically scanning lidar error in complex terrain. *Meteorol. Z.* **2009**, *18*, 189–195. [CrossRef]
13. Müller, H.; Pape, N.; Eggert, M.; Westermann, D.; Albers, A. Bedeutung laseroptischer Verfahren für die Windgeschwindigkeitsmessung. *Exp. Strömungsmech.* **2010**, *1*, 2.1–2.7.
14. Ando, T.; Kameyama, S.; Hirano, Y. All-fiber coherent Doppler LIDAR technologies at Mitsubishi Electric Corporation. *IOP Conf. Ser. Earth Environ. Sci.* **2008**, *1*, 012011 [CrossRef]
15. Harris, M.; Constant, G.; Ward, C. Continuous-wave bistatic laser Doppler wind sensor. *Appl. Opt.* **2001**, *40*, 1501–1506. [CrossRef] [PubMed]
16. Eggert, M.; Müller, H.; Többen, H. Doppler-Lidar-Transfernormals zur Windgeschwindigkeitsmessung: Aktueller Entwicklungsstand. *Exp. Strömungsmech.* **2012**, *20*, 10.1–10.6.
17. Eggert, M.; Müller, H.; Többen, H. Konzeption eines Doppler-Lidar-Transfernormals zur Windgeschwindigkeitsmessung. *Exp. Strömungsmech.* **2011**, *19*, 45.1–45.6.
18. Eggert, M.; Müller, H.; Többen, H. Doppler-Lidar-Transfernormal zur ortsaufgelösten, vektoriellen Windgeschwindigkeitsmessung. *Exp. Strömungsmech.* **2013**, *21*, 43.1–43.7.
19. Eggert, M.; Gutsmuths, C.; Müller, H.; Többen, H. Zeitaufgelöste, vektorielle Vergleichsmessungen zwischen dem Doppler-Lidar-Transfernormal der PTB und einem Referenz-Ultraschallanemometer. *Exp. Strömungsmech.* **2014**, *22*, 11.1–11.8.
20. Gutsmuths, C.; Eggert, M.; Müller, H.; Többen, H. Zeitaufgelöste, vektorielle Vergleichsmessungen zwischen dem Doppler-LIDAR-Transfernormal der PTB und konventionellen LIDAR-Systemen. *Exp. Strömungsmech.* **2015**, *23*, 65.1–65.7.

21. Eggert, M.; Gutsmuths, C.; Oertel, S.; Müller, H.; Többen, H. Untersuchungen zur Vergleichbarkeit von instantanen Strömungs-geschwindigkeitsmessungen des bistatischen Doppler-Lidars der PTB und trägheitsbehafteten Messungen eines Schalenster-nanemometers. *Exp. Strömungsmech.* **2017**, *25*, 54.1–54.7.
22. Eggert, M.; Gutsmuths, C.; Müller, H.; Többen, H. Zeitaufgelöste, vektorielle Vergleichsmessungen zwischen dem Doppler-Lidar-Transfernormal der PTB und einem 135 m hohen Windmessmasten. *Exp. Strömungsmech.* **2016**, *24*, 3.1–3.7.
23. Oertel, S.; Eggert, M.; Gutsmuths, C.; Wilhelm, P.; Müller, H.; Többen, H. Windkanalmesseinrichtung für die Validierung des bistatischen PTB-Wind-Lidars als Bezugsnormal. *Exp. Strömungsmech.* **2018**, *26*, 48.1–48.8.
24. Oertel, S.; Eggert, M.; Gutsmuths, C.; Wilhelm, P.; Müller, H.; Többen, H. Bistatic wind lidar system for traceable wind vector measurements with high spatial and temporal resolution. In Proceedings of the FLOMEKO, Lisbon, Portugal, 26–28 June 2019; Volume 18.
25. Oertel, S.; Eggert, M.; Gutsmuths, C.; Wilhelm, P.; Müller, H.; Többen, H. Validation of three-component wind lidar sensor for traceable highly resolved wind vector measurements. *J. Sens. Sens. Syst.* **2019**, *8*, 9–17. [CrossRef]
26. Kolmogorov, A.N. The local structure of turbulence in incompressible viscous fluid for very large Reynolds numbers. *CR Acad. Sci. URSS* **1941**, *30*, 301–305.
27. Kaimal, J.C.; Finnigan, J.J. *Atmospheric Boundary Layer Flows: Their Structure and Measurement*; Oxford University Press: New York, NY, USA, 1994.
28. Dabberdt, W.F. Tower-Induced Errors in Wind Profile Measurements. *J. Appl. Meteorol.* **1968**, *7*, 359–366. [CrossRef]
29. Hansen, M.O.L.; Pedersen, B.M. Influence of the Meteorology Mast on a Cup Anemometer. *ASME J. Sol. Energy Eng.* **1999**, *121*, 128–131. [CrossRef]

Article

Automated Detection of Premature Flow Transitions on Wind Turbine Blades Using Model-Based Algorithms

Ann-Marie Parrey *, Daniel Gleichauf, Michael Sorg and Andreas Fischer

Bremen Institute for Metrology, Automation and Quality Science, University of Bremen, 28359 Bremen, Germany; d.gleichauf@bimaq.de (D.G.); m.sorg@bimaq.de (M.S.); andreas.fischer@bimaq.de (A.F.)
* Correspondence: am.parrey@bimaq.de

Abstract: Defects on rotor blade leading edges of wind turbines can lead to premature laminar–turbulent transitions, whereby the turbulent boundary layer flow forms turbulence wedges. The increased area of turbulent flow around the blade is of interest here, as it can have a negative effect on the energy production of the wind turbine. Infrared thermography is an established method to visualize the transition from laminar to turbulent flow, but the contrast-to-noise ratio (CNR) of the turbulence wedges is often too low to allow a reliable wedge detection with the existing image processing techniques. To facilitate a reliable detection, a model-based algorithm is presented that uses prior knowledge about the wedge-like shape of the premature flow transition. A verification of the algorithm with simulated thermograms and a validation with measured thermograms of a rotor blade from an operating wind turbine are performed. As a result, the proposed algorithm is able to detect turbulence wedges and to determine their area down to a CNR of 2. For turbulence wedges in a recorded thermogram on a wind turbine with CNR as low as 0.2, at least 80% of the area of the turbulence wedges is detected. Thus, the model-based algorithm is proven to be a powerful tool for the detection of turbulence wedges in thermograms of rotor blades of in-service wind turbines and for determining the resulting areas of the additional turbulent flow regions with a low measurement error.

Keywords: image processing; pattern recognition; wind energy turbines; turbulence wedges

Citation: Parrey, A.-M.; Gleichauf, D.; Sorg, M.; Fischer, A. Automated Detection of Premature Flow Transitions on Wind Turbine Blades Using Model-Based Algorithms. *Appl. Sci.* **2021**, *11*, 8700. https://doi.org/10.3390/app11188700

Academic Editor: Zhengjun Liu

Received: 7 August 2021
Accepted: 14 September 2021
Published: 18 September 2021

Publisher's Note: MDPI stays neutral with regard to jurisdictional claims in published maps and institutional affiliations.

Copyright: © 2021 by the authors. Licensee MDPI, Basel, Switzerland. This article is an open access article distributed under the terms and conditions of the Creative Commons Attribution (CC BY) license (https://creativecommons.org/licenses/by/4.0/).

1. Introduction

1.1. Motivation

Electrical energy created by wind turbines has become an increasingly important part in providing clean power. However, wind turbines are exposed to many environmental influences (e.g., hail [1], rain or insects [2]) that contribute to defects such as erosion and contamination of the rotor blade, especially at the leading edge. Defects influence the geometry and surface quality of the rotor blade and may lead to premature transitions from laminar to turbulent flow in the boundary layer. Premature transitions create distinctly wedge-shaped areas of turbulent flow in otherwise laminar flow regions, the so-called turbulence wedges [3]. The increase in the overall surface area with turbulent flow due to the existence of turbulence wedges can amplify acoustic emissions [4] as well as aerodynamic imbalances [5]. Furthermore, an increase in area with turbulent flow negatively affects the aerodynamic properties (i.e., decrease in lift, increase in drag), and thus reduces the annual energy production [6]. Therefore, it is necessary to monitor the condition of the blade's surface that influences the flow to quantify the amount of additional turbulent flow area. Infrared thermography is an established contactless, in-process measurement technique to visualize laminar and turbulent flow regions on operating wind turbines without blade modifications [7–9]. A temperature difference exists between different boundary layer flow regimes due to varying local heat transfer coefficients. Using this technique, areas with turbulent flow such as the turbulence wedges can be visualized. A heating of the rotor surface is desirable to increase the temperature difference between the blade surface and fluid, which increases the contrast between laminar and turbulent flows. However, often, no active heating is available when operating wind turbines and the blade is only heated passively by sunlight. For this reason, an image processing algorithm is required which is capable of reliably detecting turbulence wedges under low contrast conditions. This would enable extensive field studies of the boundary layer flow state around the blade of operating turbines. In addition, the required algorithm has to determine the wedges' features such as position and size with a low uncertainty to finally quantify the resulting total area with turbulent flow.

1.2. State of the Art

Various methods exist for detecting laminar–turbulent flow transitions in thermograms, yet most approaches explicitly ignore turbulence wedges. One such approach called 'local infrared thermography' is presented by Mertens et al. [10] to detect unsteady transitions in periodic pitching processes. A pitching airfoil was investigated in a wind tunnel and thermograms were captured over multiple pitching periods. The airfoil was externally heated with a spotlight to increase the temperature difference between the airfoil and boundary layer flow. To detect the flow transition, the intensities of each pixel were detected over time and assigned to the corresponding phase of the pitch angle. Extrema in the intensity signal mean that the transition is passing through the current location. While the method successfully detects the flow transition, the analysis is not applicable to single thermograms, which is often the only type of data available from field measurements of in-service wind turbines. As mentioned before, the external active heating is also not often realized in field measurements.

Crawford et al. [11] located the flow transition in a crossflow-dominated environment on a heated swept wing model. The wing model was tested in wind tunnel and flight experiments, where the model was mounted on an airplane. The crossflow-dominated environment produces a jagged transition front, also called a sawtooth transition pattern. This sawtooth pattern is similar to turbulence wedges, albeit on a smaller scale. Thus, the measurement task is similar to detecting turbulence wedges. Furthermore, one transition front did in fact include one turbulence wedge. The heating of the model was realized with internal electrical heating wires. A special coating which reduced reflections, combined with the heating, improved the contrast-to-noise ratio (CNR) between laminar and turbulent flow regions in the thermograms. The saw-toothed transitions were detected through local maxima in the intensity gradient profiles after a series of image filters. Afterwards, a statistical analysis was performed to determine a prominent transition position of the sawtooth transition pattern. As the focus was put on estimating a transition position from the sawtooth pattern, turbulence wedges were intentionally rejected by the analysis. This analysis works with single thermograms but utilizes many filters and processing steps to reach its results, which have to be adjusted and optimized. Furthermore, the heating of the blade and the coating of the blade cannot be realized in field measurements of wind turbines in motion.

To detect the natural flow transition onset and end on helicopter rotor blades, Richter and Schülein [12] analyzed thermograms of rotating blades. They investigated a model blade on a whirl tower as well as full-scale rotor blades on helicopters during ground run and hovering. To increase the temperature differences between the blade surface and fluid, the blade was either spun fast to cool it, cooled with ice or passively heated in sunlight. In the thermograms, intensity profiles in the direction of the chord of the blade were examined, omitting all chord-wise positions of premature transitions. The intensity profiles consist of linear sections with different slopes. The three distinct slopes can be attributed to the laminar, transitional and turbulent regions of the boundary layer flow. Each slope was estimated with a linear function. The intersection point of the linear estimate of two adjacent regions was then classified as the transition onset and transition endpoint, respectively. The natural transition was successfully detected even for low signal-to-noise ratios, which resulted from short exposure times. However, as turbulence wedges were excluded from the analysis, premature flow transitions were not detected.

Dollinger et al. [13] investigated thermograms of wind turbines in operation to determine the flow transition. To determine the transition position with subpixel accuracy, each chord-wise intensity profile in the thermographic image was approximated with the Gaussian error function. However, turbulence wedges were not detected reliably due to the low CNR between laminar and turbulent flow regions. Similarly, Gleichauf et al. [14] used intensity gradients to also detect flow transitions in thermograms of a wind turbine in operation. Since chord-wise intensity gradients are not sufficient for the reliable detection of turbulence wedges, the thermogram was at first rotated to such a degree that the subsequently evaluated intensity gradient was perpendicular to the premature transition lines. As a result, the rotation increased the sensitivity for the flow transition detection with regard to premature flow transitions. However, turbulence wedges with low CNR to the surrounding laminar flow still remain a challenge.

The current algorithms for detecting premature transitions only make use of image information such as intensities, which are often evaluated in single pixel lines without context or comparison to neighboring image parts. Therefore, when premature transitions are detected, they account for single points in the transition front. However, neighboring premature transition points are not grouped together. Not recognizing turbulence wedges as a whole complicates the quantification of the wedges' features, such as their sizes. Only Gleichauf et al. [14] and Dollinger et al. [13] quantify figure of merits related to additional turbulent area due to premature transitions, by calculating the differences between the positions of the natural and premature transition lines instead of adding up the turbulent area of turbulence wedges. Furthermore, pattern recognition has not been utilized for turbulence wedge detection by any of the discussed approaches. Pattern recognition would allow for the detection of each turbulence wedge as a whole, based on certain features such as their shape. An algorithm which specifically uses pattern recognition to reliably detect turbulence wedges even under low CNR conditions has not been reported yet, although such an algorithm seems promising for extracting and quantifying the wedges' features such as the additional turbulent area with low uncertainty.

1.3. Aim and Outline

An automated, model-based image processing algorithm is introduced, which reliably detects premature laminar–turbulent flow transitions in thermographic flow visualization images even in low-contrast scenarios. The fact that premature transitions lead to wedge-shaped areas of turbulent flow is incorporated in the algorithm with the use of wedge-shaped templates. These templates are then used to detect turbulence wedges and also to determine the positions and sizes of the turbulence wedges in thermograms. As a result, the additional amount of area with turbulent flow originating from premature transitions can be quantified. In order to characterize the capabilities of this approach, simulations as well as validation experiments on thermograms of in-process wind turbines are performed.

Section 2 contains the description of the novel image processing algorithm as well as a definition of the area with turbulent flow resulting from premature transitions. In Section 3, the implementation of the algorithm and the simulation and measurement setup are explained. The results of the verification of the algorithm for the simulation and the validation of real thermograms from in-process wind turbine measurements are presented and discussed in Section 4. Section 5 provides concluding remarks and an outlook.

2. Measurement Approach

2.1. Thermogram Characteristics and Measurands

The measurement quantities from a thermogram of a rotor blade with a premature flow transition are the position x_i of the turbulence wedge and its size, consisting of the height h_i and the width w_i, where i is the running index of the wedge number. In Figure 1 (left), a real thermogram is shown with the marked position and size of one turbulence wedge as an example. Note that the actual temperatures of the different flow regimes are irrelevant for the wedge detection algorithm, which is why all values in the thermograms are interpreted as intensities between 0 and 1 by normalizing with the maximum intensity I_{max}.

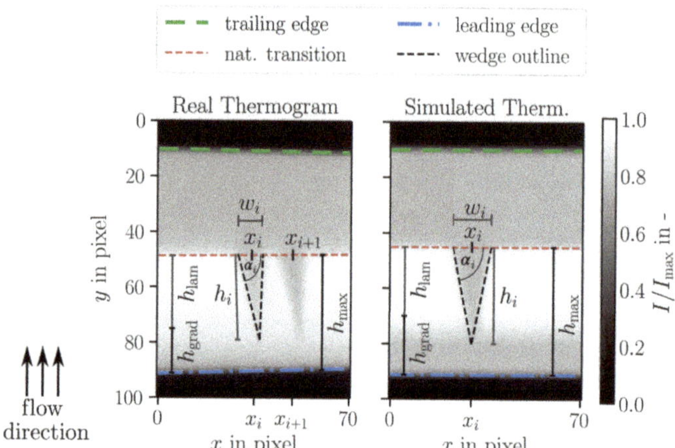

Figure 1. A real thermogram of a rotor blade of a wind turbine in operation with two turbulence wedges (**left**) and a simulated thermogram with one turbulence wedge (**right**). In this case, the rotor blade is warmer than the surrounding fluid, which means that the rotor blade is cooled by the boundary layer flow. The cooler the temperature, the lower the pixel intensity. Important characteristics of the thermograms are indicated.

According to Figure 1, the thermogram shows the leading edge and the trailing edge of the rotor blade, as well as the natural transition line that indicates the position of the non-premature flow transition from laminar to turbulent. The region of interest for the detection of turbulence wedges lies between the leading edge and the natural transition line, i.e., their distance equals the maximal height h_{max} of a turbulence wedge. In general, $N \in \mathbb{N}$ turbulence wedges are visible in one thermographic image. The height h_i of a turbulence wedge is defined as the orthogonal distance of the wedge's tip to the natural transition line, i.e., the wedge's base. The width w_i is defined as the width of the base. The position of the turbulence wedge is defined as $(x_i, y_i), i \in \{1, \ldots, N\}$, where x_i is the center of the wedge's base. The y-position of the wedge's base is set to $y_i = y_{NT}(x_i)$, where $y_{NT}(x_i)$ denotes the y-position of the natural transition line at x_i. The position x_i together with the height h_i and the width w_i of each turbulence wedge are the measurement quantities returned by the wedge detection algorithm. Furthermore, the height and the width are used to calculate the area A_i of the turbulence wedge, with which the additional turbulent area due to premature transitions is quantified.

One further characteristic feature of the turbulence wedge is its symmetry or the lateral tip position. The tips of turbulence wedges in real thermograms are often not centered below the middle of their base. An example is shown in Figure 1 (left), where the wedges are skewed to the right. The wedges can appear skewed due to the camera perspective or due to the rotational movement of the rotor blade, which implies crossflow. The tip's position can be described using the angle α_i between the left edge of the turbulence wedge and the wedge's base. While the tip's position does not influence the size of the area of the turbulence wedge, it still needs to be taken into account for the detection of the turbulence wedge with the model-based algorithm.

A detailed analysis of real thermograms revealed an intensity gradient in the region between the leading edge and natural transition line, which is characterized by decreasing pixel intensities towards the leading edge. Due to the curvature of the blade at the leading edge and the dependency of the emission on the observation direction [15], lower pixel intensities were captured in the thermogram at the leading edge. Furthermore, the heat transfer coefficient in the laminar flow regime was higher near the leading edge, where the laminar boundary layer was not yet fully developed, which also led to lower pixel intensities. The resulting intensity gradient in the thermogram begins at the leading edge and extends over a height h_{grad}. The remaining flow region to the natural transition line, with the height h_{lam} has an almost constant intensity in the thermographic image. Since the intensity gradient near the leading edge reduces the image contrast of the turbulence wedge's tip, the accurate detection of the wedge's tip is challenging.

In order to verify and characterize the functionality of a wedge-detection algorithm, a ground truth needs to be known. Therefore, simulated thermograms are necessary, which emulate the characteristic features of real thermograms for given values of the measurement quantities. A simulated thermogram is shown in Figure 1 (right), which is not an exact replication of the real thermogram to its left but models all important characteristics. It shows only a single wedge with a different angle α_i compared to the real thermogram to illustrate the relation between the angle and the tip position of the wedge. Furthermore, the intensity gradient in the simulated thermogram is exaggerated compared to the real thermogram for the sake of visibility. A detailed description of the simulation of thermograms follows in Section 3. Simulated thermograms are already utilized in the subsequent description of the functionality of the algorithm principle for the wedge detection.

2.2. Determination of the Positions of the Blade Edges and the Natural Transition Line

Turbulence wedges are located in the region between the leading edge and the natural transition line. In order to focus the turbulence wedge detection on this region, the natural transition line and the leading edge need to be located in the thermogram first. To detect the y-positions of the blade edges and the natural transition line, a chord-wise gradient approach is chosen: for each image column, which corresponds to the chord-wise y-direction, the spatial derivative $\mathrm{d}I/\mathrm{d}y$ of the pixel intensities I is calculated and then normalized with its maximum value:

$$\left(\frac{\mathrm{d}I}{\mathrm{d}y}\right)_{\text{norm}} = \left|\frac{\mathrm{d}I}{\mathrm{d}y}\right| \cdot \left(\max\left(\left|\frac{\mathrm{d}I}{\mathrm{d}y}\right|\right)\right)^{-1}. \tag{1}$$

After the normalization, all local maxima above a threshold value of $0.1\,\text{pixel}^{-1}$ are located. This threshold value is heuristically based as it suppresses noise but allows the local maxima due to the natural transition to be detected. A lower threshold value would lead to more false detections due to noise; a higher value would miss more actual transition points. To illustrate how the local maxima correspond to the y-positions of the blade edges and the natural transition, a simulated thermogram with a single turbulence wedge is shown in Figure 2a, for which the gradient is calculated for two different image columns. The studied image columns are chosen so that one (marked with a blue dotted line) intersects the turbulence wedge and therefore contains a premature transition. The other (marked with a dashed orange line) does not intersect a turbulence wedge and thus contains a natural flow transition.

According to Figure 2a, the pixel intensities of the background and the blade differ significantly, which leads to large local maxima in the gradient at the positions of the leading and the trailing edge, respectively. The natural transition also leads to a local maximum of the intensity gradient due to a large difference in pixel intensities between the laminar and the turbulent flow regions. Therefore, three local maxima can be present in the intensity gradient, corresponding to the trailing edge, the natural transition and the leading edge, see Figure 2b, at the top. The y-positions of the maxima are set as the y-positions of the blade edges and the natural transition line. In cases where only two maxima exist, see Figure 2b, at the bottom, the maxima are assigned to the trailing and leading edges. This case occurs when the chord-wise intensity profile contains a premature transition, which leads to intensity gradients that are below the threshold. Hence, the natural transition is not detectable in image columns where a premature transition exists. Therefore, the detected points of the natural transition are fitted with a linear regression so that the y-position y_{NT} of the natural transition is defined for each x-position of the thermogram and also for the region of the premature flow transition.

With the extracted y-positions of the natural transition line and the leading edge, the wedge detection algorithm can now be applied to the thermogram.

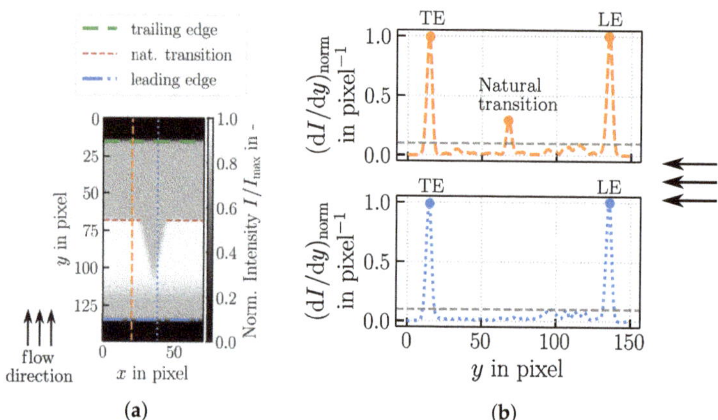

Figure 2. Simulated thermogram (**a**) and normalized gradient (**b**) for two chord-wise positions. The blue line intersects a turbulence wedge, which results in less pronounced maxima of the gradient below the threshold of $0.1\,\mathrm{pixel}^{-1}$, marked with a dashed gray line. (**a**) Simulated thermogram with one turbulence wedge. The trailing edge is labeled with a green dashed line, the leading edge with a blue dash-dotted line. The natural transition is marked with a red dashed line, excluding premature transitions. The two vertical lines mark the positions for the intensity gradients, see (**b**). (**b**) Intensity gradients normalized by their respective maximums for the two lines at $x_{\mathrm{orange}} = 20\,\mathrm{pixel}$ (**top**) and $x_{\mathrm{blue}} = 38\,\mathrm{pixel}$ (**bottom**). Local maxima are marked with dots and correspond to the trailing edge (TE), the natural transition and the leading edge (LE).

2.3. Wedge Detection Algorithm

In order to detect turbulence wedges in thermograms, a model-based algorithm is proposed, which uses a technique called template matching. Template matching finds image parts that match the used template, which means that the proper choice of the template on the basis of a priori knowledge is crucial for the functionality of the algorithm. As a preface, the wedge-shaped templates are described in Section 2.3.1. Then, the two parts of the wedge detection algorithm are explained: the detection of the turbulence wedge's position in Section 2.3.2 and the determination of the wedge's size in Section 2.3.3.

2.3.1. Wedge Template

Template matching uses a template whose shape is similar to the desired feature that is to be found in the image. Therefore, the template has a triangular shape with the three parameters height h', width w' and angle α'—see Figure 3. The angle α' is used to set the position of the template's tip $(x_{\mathrm{tip}}, y_{\mathrm{tip}})$.

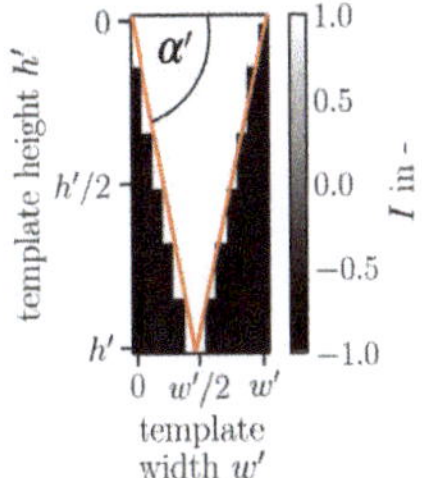

Figure 3. Example of a turbulence wedge template with height h', width w' and angle α', which determines the position of the tip.

In the coordinate system of the template, the tip's x-position can be calculated by using the equation

$$x_{\text{tip}} = h' \cdot \cos(90 - \alpha'), \tag{2}$$

where the y-position of the tip equals the wedge height, i.e.,

$$y_{\text{tip}} = h'. \tag{3}$$

Note that the width of the wedge at the base is not changed when the position of the tip is adjusted to tailor the template to the turbulence wedge in the thermogram. Currently, the adjustment of the angle α' is carried out manually, because α' can be kept constant for all turbulence wedges in the studied thermographic image.

Each template pixel value v inside the wedge was set to $v_{\text{wedge}} = 1$, while the surrounding parts were set to -1—see Figure 3. The objective of doing so is to penalize all image pixels that do not fit the wedge shape. Note that the height and the width values of the template are positive integer values, because a subpixel interpolation has not been considered yet. The described template is subsequently utilized in the wedge detection algorithm to determine the positions of the turbulence wedges as well as the areas of the turbulence wedges.

2.3.2. Detection of the Wedge Position

In order to detect the positions of the turbulence wedges, a cross-correlation of the thermogram with the wedge template is used:

$$C[x] = \sum_{p} \sum_{q} \iota_{p,q} \cdot \tau_{p-y_{\text{NT}}(x),\, q-x'} \tag{4}$$

where τ denotes the template and ι denotes the thermogram section with the same size of the template. The variable p is the row index of each image matrix, while q is the column index. In accordance with the natural occurrence of turbulence wedges on a rotor blade, the template's y-position equals the y-position $y_{\text{NT}}(x)$ of the natural transition line, leading to a single cross-correlation result $C[x]$ for each x-position. The process of the detection of the wedges' positions is exemplified in Figure 4 using a simulated thermogram with three wedges, shown in Figure 4a.

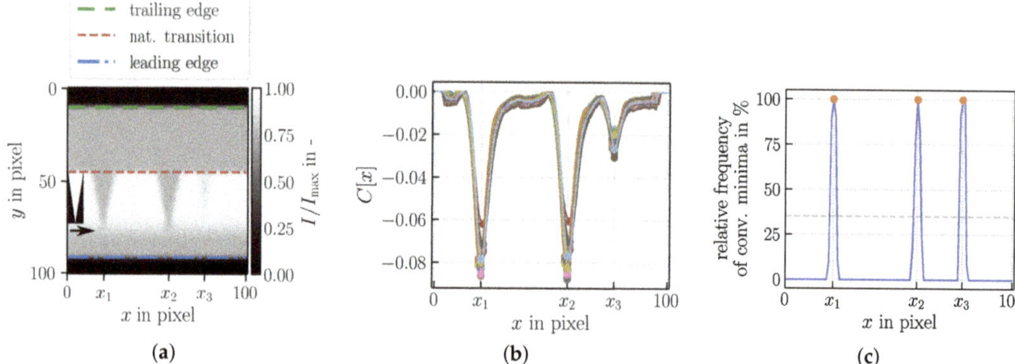

Figure 4. Illustrative analysis of a simulated thermogram (**a**) to detect the turbulence wedges' positions. The position detection is implemented using cross-correlations with randomized templates (**b**). Maxima in the relative frequency of minima in the cross-correlation results (**c**) determine the position of the turbulence wedges. (**a**) Simulated thermogram with three differently sized wedges. The wedge on the right has a lower contrast. The rectangle on the left side represents a template in its starting position, with an arrow below indicating the direction in which it is moved pixel-wise during the cross-correlation. (**b**) Cross-correlation $C[x]$, see Equation (4), of the simulated thermogram in (**a**) and $J = 100$ templates of varying sizes. Each color corresponds to a template of a different size. In this example, the smallest template has a size (h', w') of $(27, 8)$ pixel; the largest $(41, 16)$ pixel. Local minima are marked with a dot. (**c**) Relative frequency of minima in the results of the cross-correlations. Each position x_i of a maximum (orange dot) is marked. The cut-off percentage of 35% is marked with a dashed gray line.

A local minimum in the results of the cross-correlation indicates a good agreement between the thermogram and the template. To enable the detection of turbulence wedges with different, yet unknown sizes, the cross-correlation calculation is repeated for $J \in \mathbb{N}$ templates with different wedge heights and widths. Each template size corresponds to a different curve in the cross-correlation results shown in Figure 4b. For each x-position, the number of local minima (marked as dots) in the cross-correlation results is counted and normalized by the total number of studied templates, which results in a relative frequency of minima for each x-position, see Figure 4c. If the relative frequency is larger than 35% at a certain x-position, which is an empirically based value, it is considered to be a likely position for a wedge. The so determined x-positions $x_i, i \in \{1, \ldots, N\}$, correspond to the middle of the wedges' bases and thus to the wanted x-position of each turbulence wedge.

In addition, at each of the found position x_i, the minimum of all cross-correlation values is determined. The associated template τ_0 with the height $h_{0,i}$ and a width $w_{0,i}$ has the best fit to the underlying wedge. The found wedge positions x_i as well as the best fitting template $\tau_{0,i}$ for each x_i position with $(h_{0,i}, w_{0,i})$ are needed for the precise determination of the wedges' areas, which is the next step of the algorithm.

2.3.3. Determination of the Wedge Area

In order to determine the additional area with turbulent flow in a thermogram caused by premature flow transitions, the area A of all N wedges in a thermogram needs to be known. The total area A is determined by summing up the area A_i of each wedge:

$$A = \sum_{i=1}^{N} A_i. \tag{5}$$

The area A_i of each turbulence wedge in the thermogram is calculated with the formula

$$A_i = \frac{1}{2} h_i w_i, \tag{6}$$

where h_i denotes the height and w_i the width of the wedge at the position x_i. In order to determine h_i and w_i, a measure of similarity between a template at the position x_i and the thermogram section of the same size is required, which is calculated for different template sizes but at a fixed position to finally identify the most fitting template size.

The chosen measure of similarity is the weighted correlation, which is selected instead of the regular correlation due to the intensity gradient near the leading edge in the thermogram, which leads to a low contrast between the wedge's tip and the surrounding region. This low contrast between tip and surrounding causes the wedge height to be estimated too small and therefore leads to an erroneously determined wedge area. By using the weighted correlation, which weights a good fit between the template and the wedge at the base stronger than at the tip, the intensity gradient near the wedge's tip is counteracted. The weighted correlation in between a template τ and the thermogram section ι is defined as

$$\text{corr}(\tau, \iota) = \frac{\text{cov}(\tau, \iota)}{\sqrt{\text{cov}(\tau, \tau)\text{cov}(\iota, \iota)}}, \tag{7}$$

the covariance is the weighted covariance according to

$$\text{cov}(\tau, \iota) = \frac{\sum_p \sum_q \gamma_{pq} \cdot (\tau_{pq} - \overline{\tau})(\iota_{pq} - \overline{\iota})}{\sum_p \sum_q \gamma_{pq}}, \tag{8}$$

with the weight matrix γ and the weighted mean values

$$\overline{\tau} = \frac{\sum_p \sum_q \gamma_{pq} \tau_{pq}}{\sum_p \sum_q \gamma_{pq}}, \quad \overline{\iota} = \frac{\sum_p \sum_q \gamma_{pq} \iota_{pq}}{\sum_p \sum_q \gamma_{pq}}. \tag{9}$$

The index p indicates the row index of each matrix, while q is the column index.

To determine the height h_i and the width w_i of the wedge at the position x_i, the template $\tau_{0,i}$ with the height $h' = h_{0,i}$ and the width $w' = w_{0,i}$, which was determined during the detection of the wedges' positions in Section 2.3.2, is used as an initial wedge size for the search. Since the sizes of the templates used in the position detection are chosen randomly, it is not ensured that $h_{0,i}$ and $w_{0,i}$ are the actual height and width of the turbulence wedge at the position x_i. Therefore, additional templates with sizes in the vicinity of the initial template size are investigated by calculating the weighted correlation according to Equations (7)–(9). The maximum value of these correlations yields the optimal size parameters (h_i, w_i) of the wedge at the position x_i, which are then inserted into Equation (6) to obtain the wedge area.

3. Implementation of the Model-Based Algorithm and the Experimental Setup

3.1. Numerical Implementation of the Algorithm

The model-based wedge detection algorithm is implemented in the programming language *Python*. The steps to determine the position x_i, the size parameters (h_i, w_i) and the area A_i of each turbulence wedge are outlined in a flowchart in Figure 5.

As a first step, either the simulated thermogram is created or the recorded thermogram of the wind turbine blade is loaded. The loaded thermogram is then preprocessed, which includes an optional normalization of the thermogram intensities by the maximum intensity as well as an image rotation. The rotation aligns the natural transition line of the thermogram horizontally, which simplifies the calculation of the correlations as the y-positions of the line remain constant. To determine the rotation angle, the y-positions of the natural transition line y_{NT} are at first detected in the unprocessed thermogram, using the classical gradient-based method described in Section 2.2. Using the coordinates of the natural transition line, the rotation angle φ is calculated according to

$$\varphi = \arctan\left(\frac{y_{\text{NT}}(x_{\text{end}}) - y_{\text{NT}}(x_0)}{x_{\text{end}} - x_0}\right), \tag{10}$$

where $y^{\text{NT}}(x_{\text{end}})$ and x_{end} are the last coordinates of the natural transition line and $y^{\text{NT}}(x_0)$ and x_0 are the first coordinates of the natural transition line in the original, i.e., non-rotated thermogram. The whole thermogram is rotated by the angle φ using the *rotate* functionality from the *Python Image Library* (PIL). Note that the image rotation is not necessary for simulated images, where the natural transition line is already aligned horizontally.

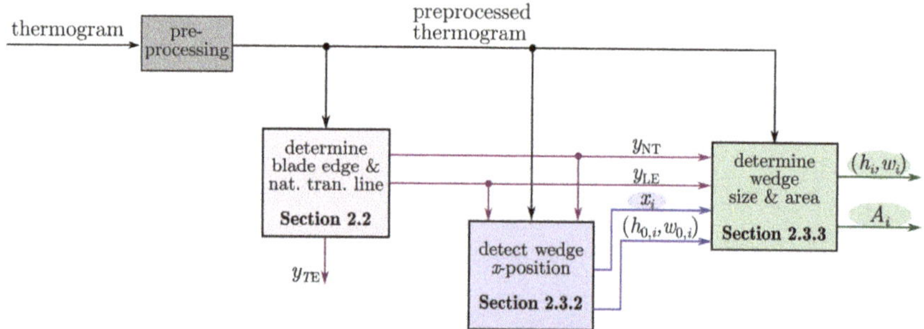

Figure 5. Flowchart of the model-based algorithm. y_{TE} are the y-positions of the trailing edge, y_{NT} the y-positions of the natural transition line and y_{LE} the leading edge's y-positions. The output quantities, which are highlighted with a colored circular background, are the x-position x_i, the height h_i, the width w_i and the area A_i of each turbulence wedge, where i denotes a running wedge number.

The preprocessing also includes the determination of the angle α_i of the turbulence wedges, which is currently the only manually estimated input parameter for the wedge detection algorithm. However, α_i remains the same for all turbulence wedges until the flow condition changes at which the thermograms are captured.

After the preprocessing, the y-positions of the leading edge y_{LE} and the trailing edge y_{TE}, as well as the natural transition line y_{NT}, are detected in the thermogram by using the gradient-based method introduced in Section 2.2. The values obtained by the gradient-based method are fitted with a linear regression using the *RANSAC* algorithm of *Python's sklearn* library. To use the leading edge and the natural transition line in the wedge detection algorithm, the y-position values are rounded to the next integer. With the y-positions of the leading edge and the natural transition line, the maximum height h_{max} of the turbulence wedges is calculated according to

$$h_{\text{max}} = \langle y_{\text{LE}}(x) - y_{\text{NT}}(x) \rangle, \tag{11}$$

where the mean is taken over all x-positions.

As the next step, the wedges and their x-positions x_i are detected along the natural transition line by cross-correlating the thermogram with the wedge-shaped templates, see Section 2.3.2. For the position detection, $J \in \mathbb{N}$ differently sized templates with running template index j are randomly generated, where $J = 100$ is the default value. The templates' heights h'_j are drawn from a uniform distribution in the interval $[0.5 \cdot h_{\text{max}}, 0.95 \cdot h_{\text{max}}]$. This interval has proven adequate for detecting wedges in real thermograms, but the interval can also be adapted to any values in $(0, h_{\text{max}}]$ if needed. The investigation of real thermograms (see Section 3.2) has further shown that the height-to-width ratio of real turbulence wedges is, on average, $\langle h_i / w_i \rangle = 3$. Consequently, the templates' widths are drawn from a normal distribution with $\mathcal{N}(\mu_{w'} = h'_j/3, \sigma^2_{w'} = 0.2)$, where $\mu_{w'}$ is the mean and $\sigma^2_{w'}$ is the variance of the normal distribution.

To calculate the cross-correlation between the thermogram and each turbulence wedge template, the template with index j is placed with its top left corner at $x_0 = 0$ and $y = y_{\text{NT}}(x_0)$ as a starting point. A section of the full thermogram the same size as the template is required for the cross-correlation calculation. The columns (i.e., the width) of the thermogram section are chosen from the position x to $(x + w'_j)$ of the full thermogram. The rows (i.e., the height) of the thermogram section are determined from the position

$y_{NT}(x)$ to $(y_{NT}(x) + h'_j)$. The cross-correlation of the thermogram section and the template is calculated using Equation (4) and the template is moved to the right by 1 pixel. This way, the cross-correlation value $C[x]$ is calculated for each x-position. However, when the right edge of the template reaches the right edge of the thermogram, the top left corner is at $x = w_{therm} - w'_j$, where w_{therm} is the width of the thermogram. For the remaining x-positions the cross-correlation can not be calculated as the template would otherwise protrude outside the borders of the thermogram. Therefore, due to the different widths w'_j of each template, the length of the cross-correlation curve of each template is different. To align the results of the cross-correlations of different templates, the cross-correlation values for each template j are shifted to the right by $w'_j/2$ and the maximum cross-correlation value is subtracted. Using the aligned cross-correlation curves, the relative frequency of minima is determined. The positions of the local maxima above 35% in the relative frequency then correspond to the wedge positions x_i. The threshold value of 35% can be adapted depending on the desired sensitivity of the algorithm. The higher the threshold value, the more contrast the turbulence wedges need to have in order to be detected, which can lead to overlooked turbulence wedges. A lower threshold value, on the other hand, can result in an erroneous detection of noisy structures in the image as a turbulence wedge. Note that the relative frequency is currently not interpolated, i.e., the wedge positions x_i are integer values.

As the last step of the wedge detection algorithm, the wedges' sizes (h_i, w_i) and their areas A_i are determined for each wedge position x_i using a weighted correlation, see Section 2.3.3. For each x_i, the initial template $\tau_{0,i}$ with the size $(h_{0,i}, w_{0,i})$ is used to create a range of new templates. Each template is then compared to the wedge to find the template with the best fit to the turbulence wedge. The sizes of the new templates are taken from a range of

$$h_{0,i} - 3\,\text{pixels} \leq h'_k \leq h_{0,i} + 3\,\text{pixels} \tag{12}$$

for the height and

$$w_{0,i} - 3\,\text{pixels} \leq w'_k \leq w_{0,i} + 3\,\text{pixels} \tag{13}$$

for the width. For each height and width combination of the two intervals, a new template is created, which results in $K = 49$ templates in total with running template index k. Using each template, the weighted correlation with the turbulence wedge at the position x_i is calculated. To calculate the weighted correlation, the top left corner of the current template matrix is placed at $(x_i - w'_k/2)$ along the natural transition line at $y_i = y_{NT}(x_i)$, resulting in the top right corner of the template to be positioned at $(x_i + w'_k/2)$. The columns (i.e., the width) of the thermogram section are selected from the x-position $(x_i - w'_k/2)$ of the thermogram to $(x_i + w'_k/2)$. The rows (i.e., the height) of the thermogram section are determined from the y-position $y_{NT}(x_i)$ to $y_{NT}(x_i) + h'_k$. Thus, the thermogram section and the template have the same size. The weighted correlation is chosen as a measure of similarity so that the intensity gradient near the leading edge of the thermogram is counteracted, which is implemented through the weights γ in the weight matrix. Thus, the weighting emphasizes a good match at the base of the template to the turbulence wedge. Therefore, the weights are larger near the base of the turbulence wedge, with $\gamma_{base} = 10$, and linearly decrease row by row to smaller weights near the tip, $\gamma_{tip} = 1$. The number of steps for the linear decrease, which was implemented with the *linspace* function of *Python's numpy*, is equal to the height h'_j of the current template. The linear decrease is repeated in each matrix column for w'_j columns. The weight matrix, the thermogram section and the template therefore all have the same size, and the weighted correlation is calculated using Equation (7). The calculations result in a phase space of correlation values, i.e., one value for each height-width combination, see Figure 6. The size of the template which attains the highest correlation value, see red x in Figure 6, is the size (h_i, w_i) of the turbulence wedge at position x_i, from which the area A_i can be calculated with Equation (6).

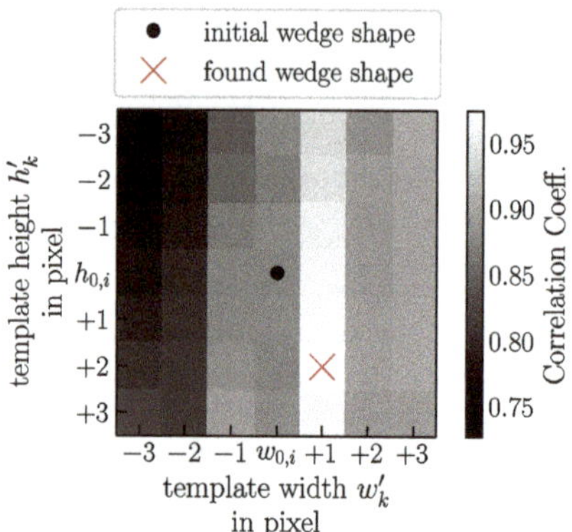

Figure 6. Phase space of correlations for a wedge at position x_i, over the template height h'_k and width w'_k. The height $h_{0,i}$ and the width $w_{0,i}$ are the initial values for the determination of the wedge's size. The phase space point $(h_{0,i}, w_{0,i})$ is marked with a black dot. The point (h'_k, w'_k) with the highest correlation value is marked with a red x.

3.2. Simulation Setup

In order to verify the wedge detection algorithm and characterize it with respect to the measurement uncertainties, simulated thermograms are needed, where all measurands of the thermogram are a priori known exactly.

The simulated thermogram consists of a matrix with a height h_{therm} and a width w_{therm}. The trailing edge is placed at $y_{\text{TE}} = 0.10 \cdot h_{\text{therm}}$, the natural transition at $y_{\text{NT}} = 0.45 \cdot h_{\text{therm}}$ and the leading edge at $y_{\text{LE}} = 0.90 \cdot h_{\text{therm}}$. The blade edges and natural transition lines are aligned horizontally. The pixel intensities of the turbulent region I_{turb} and the laminar region I_{lam} are quantities that need to be modelled to values in real thermograms. The pixel intensity of the background is set to 0.1, as a low value results in a strong gradient to the blade edge. However, the pixel intensity of the background does not influence the wedge detection algorithm and the exact value is therefore inconsequential.

Moreover, the intensity gradient near the leading edge is simulated. Near the natural transition line, the laminar value is prevalent. Towards the leading edge, the pixel intensities start to linearly decrease at some y-position $y_{\text{LE}} - h_{\text{grad}}$, see Figure 1 (right). Then, at a y-position $y_{\text{LE}} - h_{\text{grad},0}$ near the leading edge, the decreasing intensity reaches the turbulent intensity value I_{turb}. After this y-position up to the leading edge, the pixel intensities are set to I_{turb}. The parameters h_{grad} and $h_{\text{grad},0}$ are modelled according to real thermograms. Note that both parameters are subsequently presented normalized by h_{max} to make them transferable to thermograms of any size.

In addition, the amount of Gaussian blur in the thermogram can be set, with a standard deviation σ_{blur} of the Gaussian kernel of 1.0 pixel. The blur value was estimated manually to match the blur in real thermograms.

Furthermore, the noise in real thermograms is simulated by creating a matrix of noise values with the same size as the thermogram, for which each pixel value is drawn from a normal distribution $\mathcal{N}(\mu_{\text{noise}}, \sigma^2_{\text{noise}})$ with mean $\mu_{\text{noise}} = 0$ and variance σ^2_{noise} in arbitrary pixel intensity units. By changing the variance σ^2_{noise}, the amount of noise in the image can be varied. The matrix is then added to the thermogram to obtain a noisy thermogram.

To emulate a turbulence wedge realistically, the parameters additionally required from real thermograms are the typical range of the height h_i and the width w_i of the turbulence wedge. Instead of the absolute height value (in pixels), the relative height h_i/h_{max} is considered here, which makes the value applicable to thermograms of any size. To emulate the average width of the turbulence wedges, the height-to-width ratio h_i/w_i is further considered. The angle α_i of each wedges is set, by default, so that the tip position is centered below the middle of the wedge for all turbulence wedges. Finally, the typical contrast-to-noise ratio (CNR) of a turbulence wedge is emulated, where the CNR is defined as

$$\mathrm{CNR_i} = \sqrt{\frac{(\langle I_{\mathrm{lam}}\rangle_i - \langle I_{\mathrm{wedge}}\rangle_i)^2}{\sigma(I_{\mathrm{lam}})_i^2 + \sigma(I_{\mathrm{wedge}})_i^2}}, \tag{14}$$

where $\langle I_{\mathrm{wedge}}\rangle_i$ and $\langle I_{\mathrm{lam}}\rangle_i$ are the mean pixel intensities in the turbulent region inside the wedge and the laminar region right outside of the turbulence wedge with index i, respectively. Furthermore, $\sigma(I_{\mathrm{wedge}})_i$ and $\sigma(I_{\mathrm{lam}})_i$ denotes the spatial standard deviation of the pixel intensities of the turbulence wedge or its surrounding area.

To realistically emulate real thermograms, the above mentioned features of real thermograms and turbulence wedges need to be quantified. For this reason, real thermograms of wind turbines in operation are analyzed. In the analysis, 14 different thermograms and 43 different turbulence wedges are included. All heights and widths are measured using the program ImageJ [16], while the CNR calculations are performed in Python. The loaded thermograms are normalized to a range of $[0, 1]$. The results can be seen in Table 1, where the first section contains the values that are important for the modelling of a thermogram without turbulence wedges. The second section contains the characteristic values of the turbulence wedges.

Table 1. Important characteristics of the thermograms of rotor blades (first section) and the turbulent wedges (second section) determined in real thermograms of wind turbines in motion. All thermograms were normalized to a range of 0 to 1 in arbitrary intensity units. The standard deviation of the mean is given as the measurement uncertainty.

Characteristic	Determined Value	Simulated Value
$\langle I_{\mathrm{lam}}\rangle$	0.96 ± 0.01	0.96
$\langle I_{\mathrm{turb}}\rangle$	0.75 ± 0.02	0.75
$\langle \sigma(I_{\mathrm{lam}})\rangle$	0.010 ± 0.001	0.009
$\langle \sigma(I_{\mathrm{turb}})\rangle$	0.009 ± 0.001	0.009
$\langle h_{\mathrm{grad}}/h_{\mathrm{max}}\rangle$	0.50 ± 0.07	0.50
$\langle h_{\mathrm{grad},0}/h_{\mathrm{max}}\rangle$	0.16 ± 0.04	0.15
$\langle h_i/h_{\mathrm{max}}\rangle$	0.73 ± 0.02	$[0.6, 0.85]$
$\langle h_i/w_i\rangle$	3.01 ± 0.07	3
$\langle \mathrm{CNR_i}\rangle$	10.1 ± 1.1	$[2, 20]$

By adapting the mean values in Table 1 in the simulated thermograms, thermograms of rotor blades of in-service wind turbines are emulated realistically. The values used in the simulation can be seen in the rightmost column. Note that the values for $\langle \sigma(I_{\mathrm{turb}})\rangle$ and $\langle \sigma(I_{\mathrm{turb}})\rangle$ are identical in the simulation, because the marginal difference in the experimental results is neglected. Furthermore, the values of the heights h_i/h_{max} of the simulated turbulence wedges are drawn from a uniform distribution in the interval $[0.6, 0.85]$ if not explicitly stated otherwise, which places the found mean value approximately in the middle of the range. The widths w_i of the simulated turbulence wedges are drawn from a normal distribution $\mathcal{N}(\mu_w = h_i/3, \sigma_w^2 = 0.2)$, where μ_w is the mean and σ_w^2 is the variance of the normal distribution. By modifying the size of the turbulence wedges in the simulated thermograms in a small interval which is easily detectable by the algorithm, no size that is particularly well or poorly detected is chosen by chance.

In order to characterize the dependency of the model-based wedge detection algorithm on the CNR value as well as the size of the turbulence wedges, Monte Carlo simulations are performed which utilize the simulated thermograms. With the results of the Monte Carlo simulations, the systematic and the random errors of the wedge position x_i, the wedge size (h_i, w_i) and the wedge area A_i are investigated.

For the investigation of the dependency of the algorithm on the CNR, the CNR of a single turbulence wedge is changed in a range of 2 to 20 by changing the pixel intensity of the turbulence wedge. The choice of the CNR range places the mean value of $\langle \text{CNR} \rangle = 10$ found in real thermograms in the middle of the range, but also covers more extreme CNR values. Varying the intensity of the turbulence wedges instead of changing the amount of noise in the image is more realistic, as the noise in real thermograms stays approximately constant throughout the image.

To investigate the dependency of the algorithm on the size of the turbulence wedges, the height of the turbulence wedge in the thermogram is changed systematically in a range of $0.1 \leq h_i/h_{\max} \leq 1.0$. Due to the identified relationship $\langle h_i/w_i \rangle = 3$, the wedge width is changed accordingly. The CNR value of the turbulence wedge, however, remains constant at $\text{CNR} = 11$ throughout this analysis.

3.3. Measurement Setup

Real thermograms of rotor blades of in-service wind turbines are required to validate the algorithm. Therefore, thermographic measurements are performed on a 1.5 MW wind turbine of the type GE 1.5 sl, manufactured by General Electric (Boston, MA, USA), with a hub height of 62 m and rotor diameter of 77 m. The thermograms are taken with an actively cooled infrared camera called imageIR 8300, manufactured by InfraTec GmbH (Dresden, Germany). This thermographic camera has an InSb focal plane array with a format of $(640 \times 512)\,\text{pixel}^2$ where 1 pixel $\widehat{=}$ 15 µm, and is sensitive to light of a wavelength of 2–5 µm. The dynamic range is 14 bit, the integration time is set to 1600 µs and the noise equivalent temperature difference is about 25 mK at 30 °C. The measurements are taken over multiple days on a wind turbines at a measurement distance of 100 m. Due to the length of the rotor blades and to improve the spatial resolution, a 200 mm telephoto lens is used. Consequently, the rotor blades are captured in segments. For this purpose, the thermographic camera is triggered externally with an optical trigger camera when the rotor blade is positioned horizontally, i.e., parallel to the ground. An example of a typical measurement setup can be seen in Figure 7, where thermograms of the suction side of the rotor blade are acquired.

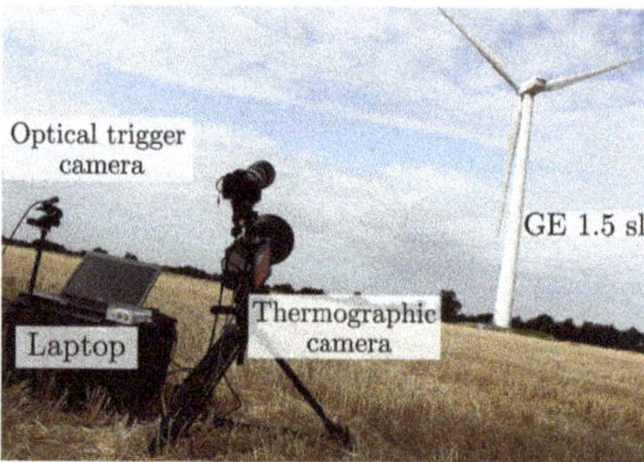

Figure 7. Photo of experimental setup for field measurements of wind turbines in motion. The thermographic camera, which takes thermograms of the wind turbine in the background, the optical trigger camera, which triggers the thermographic camera, and the laptop, which acquires the images, can be seen.

With the measurement setup, thermographic measurements of different rotor blade segments are taken, where the turbulence wedge have varying sizes as well as different CNR values, which are used to validate the wedge detection algorithm. Furthermore, the distance between wedges differs between the measurements, which has not been considered in the verification and which demonstrates the applicability of the wedge detection algorithm to real thermograms.

4. Results

In this section, the results of the image processing algorithm for turbulence wedge detection are presented. First, a verification and detailed characterization of the algorithm on simulated thermograms is performed in Section 4.1 and Section 4.2, respectively. Particularly, the dependency of the algorithm on the CNR value and the size of the turbulence wedge is investigated. The validation of the algorithm for measured thermograms of wind turbine blades in operation follows in Section 4.3. Note that the results of the verification and the validation with the model-based algorithm (MBA) are compared with the state-of-the-art, gradient-based image processing algorithm (GBA) from [14].

4.1. Verification

A verification of the model-based wedge detection algorithm is performed with a simulated thermogram with known position and size of the wedges. The studied thermogram contains three wedges, I, II and III, with the CNR values $19, 5$ and 3. The simulated thermogram with the consecutively numbered turbulence wedges of different sizes is shown in Figure 8a, while the reference turbulence wedge outline as well as the CNR value are shown in Figure 8b. Both of the wedge detection algorithms result in a flow transition line which includes all premature transitions across the width of the thermogram, see Figure 8d. To determine the area of the turbulence wedges with the GBA, the y-coordinates of the natural transition line are subtracted from the y-coordinates of the detected premature transition line. The results of the GBA and the novel MBA are shown in Figure 8c,d, respectively. The GBA detects wedge I, but fails to detect the other two wedges with lower CNR values. In contrast, the MBA correctly detects all three wedges. The results of the verification thus show that, with the introduced model-based algorithm, a reliable detection even at $CNR = 3$ is possible, which is a significant improvement compared to the state-of-the-art, gradient-based algorithm.

A quantitative comparison of the areas found by the two wedge detection algorithms is shown in Table 2. The total area determined by the MBA has a relative deviation of 1.5% compared to the true value, while the state-of-the-art GBA detects a total area with a relative deviation of -38.6% from the true value due to the undetected turbulence wedges. The MBA determines a larger-than-true area for the smallest turbulence wedge in the middle, but smaller areas for the larger turbulence wedges, which seems to be worth investigating further in the subsequent, more detailed characterization.

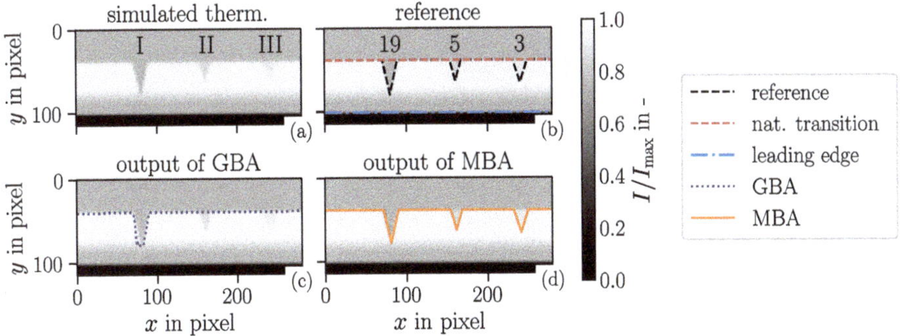

Figure 8. Simulated thermogram with three different turbulence wedges with differing intensities, leading to different CNR values. Both the model-based algorithm as well as the state-of-the-art, gradient-based algorithm were applied to the image. (**a**) shows the simulated thermogram with the turbulence wedges numbered with roman numerals. (**b**) shows the reference transition line as well as the CNR value of the wedges noted above the respective wedge. (**c**) shows the result of applying the gradient-based algorithm (GBA) of Gleichauf et al. [14] to the image and (**d**) shows the result of the application of the model-based algorithm (MBA) on the simulated thermogram.

Table 2. Areas determined by the model-based algorithm (MBA) as well as the state-of-the-art, gradient-based algorithm (GBA) for the wedges in Figure 8. The area values are given in unit pixel^2.

	True Value	GBA	MBA
Wedge I	389.5	512.0	370.5
Wedge II	187.5	-	180.0
Wedge III	256.5	-	270.0
Total	833.5	512.0	846.5
Measurement error normalized by the true value in %	0	−38.6	1.5

4.2. Characterization

To characterize the model-based algorithm in more detail, Monte Carlo simulations with thermograms of the size (140×140) pixels^2 are utilized, with which the dependency of the model-based algorithm on the CNR value as well as on the size of the turbulence wedges is examined. The deviations of the detected number N of turbulence wedges, the measured position x_i and the size parameters (h_i, w_i) as well as the area A_i are examined.

First, the dependency of the measurement results on the CNR value is investigated. The CNR value is varied in a range of $2 \leq \text{CNR} \leq 20$ by changing the pixel intensity of the $\tilde{N} = 1$ turbulence wedge present in the simulated thermogram. For each CNR value, $n = 100$ thermograms are investigated. Note that the other parameters used in the simulation, such as the wedge size and the intensity gradient, are stated in Section 3.2.

The number N of turbulence wedges that are detected in a thermogram is assessed and compared to the true value $\tilde{N}$. Here, the relative error

$$F_N = \frac{N_i - \tilde{N}_i}{\tilde{N}_i} \tag{15}$$

is investigated, where N_i is the number of detected wedges in the ith thermogram and $\tilde{N}_i = 1$ is the true number of wedges. In Figure 9, the systematic error $\langle F_N \rangle$ is shown as a function of the CNR. Only for the lowest CNR value (CNR = 2), the number of detected wedges is not equal to the true number of wedges, where 7% of turbulence wedges are missed.

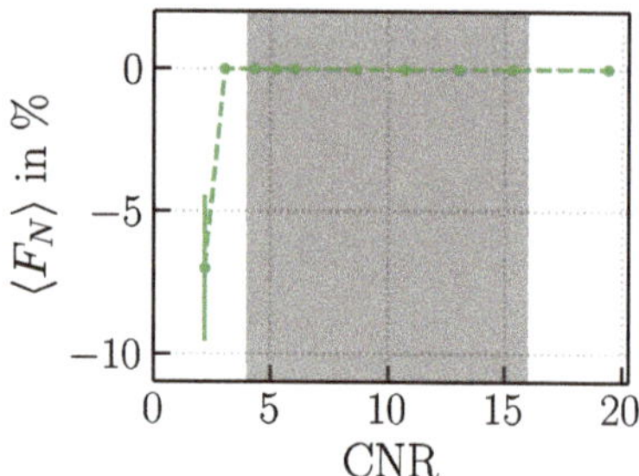

Figure 9. Mean relative error $\langle F_N \rangle$ for different CNR values. For each CNR value, $n = 100$ thermograms were investigated. The gray region marks CNR values found in real thermograms, CNR $= 10 \pm 1\,\sigma$, with $\sigma = 6$. The error bar indicates the standard deviation of the mean.

In all other cases, $\langle F_N \rangle = 0$, which means that the number of found wedges is equivalent to the true number of wedges in the thermogram, i.e., the detection error rate is zero. Furthermore, the position error Δx normalized by the average true width $\langle \tilde{w} \rangle$ of the wedges, is investigated:

$$\frac{\Delta x}{\langle \tilde{w} \rangle} = \frac{1}{\langle \tilde{w} \rangle}(x_i - \tilde{x}_i). \tag{16}$$

Here, x_i is the x-position of the found wedge in the ith thermogram, $\tilde{x}_i$ is the true x-position and $\langle \tilde{w} \rangle = 15.01$ pixels is the average true width of all wedges used. Only the

x-position needs to be considered as the algorithm detects the wedges along the natural transition line, i.e., the y-position follows from the x-position and the determined natural transition line. Figure 10 shows $\frac{\Delta x}{\langle w \rangle}$ as a function of the CNR. For the lowest CNR value of CNR $= 2$, seven thermograms were excluded from the analysis as no turbulence wedge was found and, therefore, x_i is unknown. The mean relative error $\left\langle \frac{\Delta x}{\langle w \rangle} \right\rangle$ stays between 2% and 4% for all CNR. The standard deviation of $\frac{\Delta x}{\langle w \rangle}$ in Figure 10 (right), which represents the random error, is larger than the systematic error for most CNR values by a factor of 1.5. However, for CNR < 3, the systematic error is larger than the random error. Therefore, the systematic error plays a significant role at low CNR < 3, where it should be corrected.

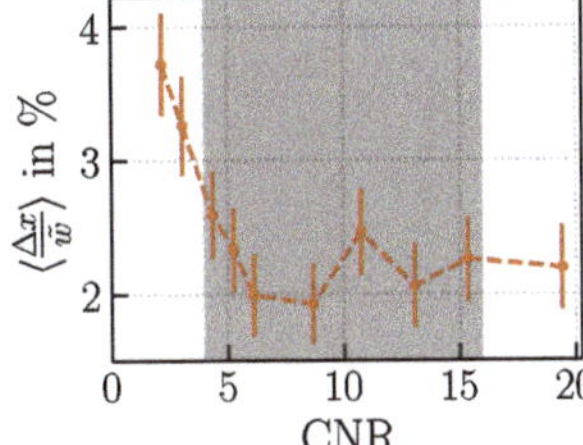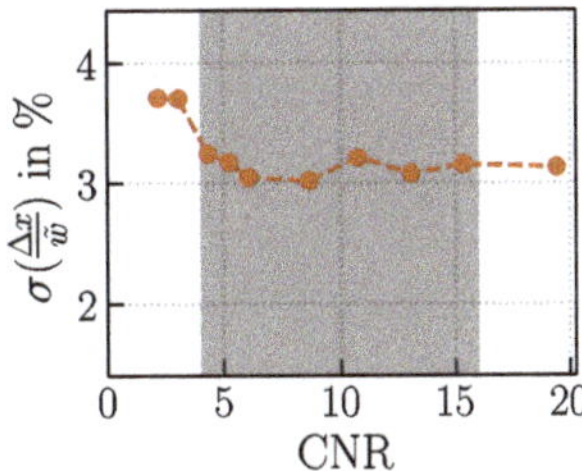

Figure 10. (**Left**): Mean position error $\left\langle \frac{\Delta x}{\langle w \rangle} \right\rangle$ normalized by the average wedge width (see Equation (16)) for $n = 100$ thermograms for different CNR values. The error bar indicates the standard deviation of the mean. The gray region marks realistic CNR values commonly found in real thermograms. (**Right**): The standard deviation of $\frac{\Delta x}{\langle w \rangle}$ in percent, which represents the random error.

To investigate the systematic and random error concerning the size of the turbulence wedge, the relative deviation of the height, the width and the area of the wedge are assessed. The relative deviation for a variable X is defined as

$$F_X = \frac{X_i - \tilde{X}_i}{\tilde{X}_i}, \tag{17}$$

where X can be the height h, the width w or the area A to define the average relative deviations F_h, F_w and F_A. Figure 11 shows the mean relative deviations of the heights (Figure 11a, left), the widths (Figures 11b, left) and the areas (Figure 11c, left) over different CNR values. In general, the lower the CNR, the larger the deviations. Furthermore, while the height of the wedge is consistently underestimated, the width is overestimated for all CNR values. The largest deviations occur at CNR < 4, which is a CNR value not often found in real thermograms as it is outside the 1σ interval. As a result, the turbulence wedge areas are underestimated throughout the investigated CNR range. The plots on the right hand side of each figure show the standard deviation of the relative deviation, which is a measure of the random error. For the width, the random error is slightly larger than the systematic error by a factor of about 1.5 for CNR > 4. However, for the height and the area, the random error is of about the same magnitude as the systematic error. Therefore, the results should be corrected by the systematic error in the future. Currently, the height and area of the wedge are underestimated on average by maximally about 10% for CNR values that are common in thermograms, $4 \leq$ CNR ≤ 16.

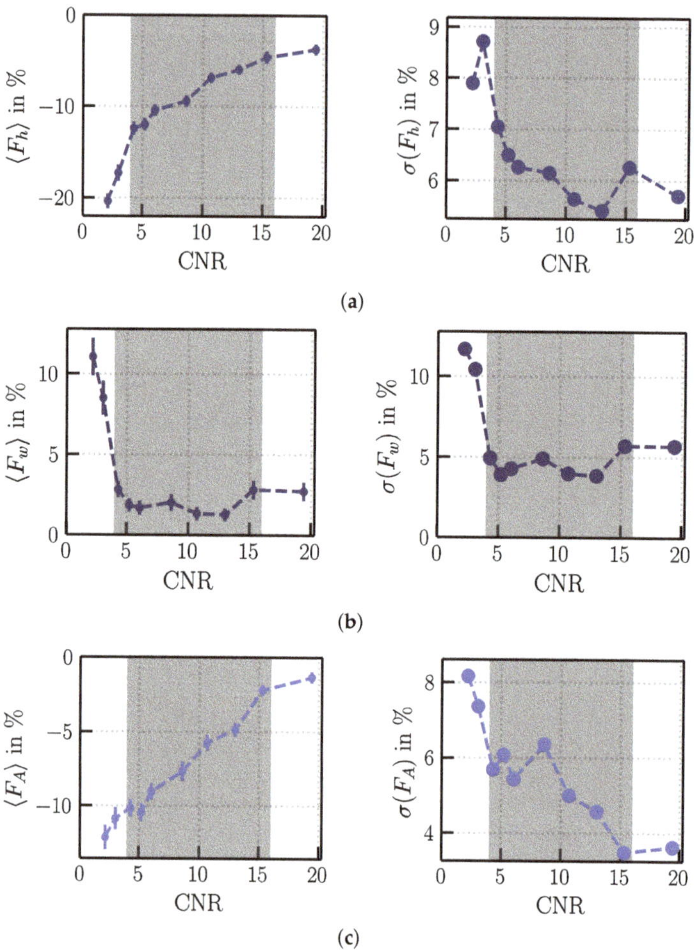

Figure 11. Left: Mean relative deviation of (**a**) the wedges' heights $\langle F_h \rangle$, the wedges' widths $\langle F_w \rangle$ (**b**) and wedges' areas $\langle F_A \rangle$ (**c**) for each CNR value with $n = 100$ trials with randomly sized wedges. The error bar indicates the standard deviation of the mean, and the gray region shows realistic CNR values found in real thermograms. **Right**: Standard deviation of all respective mean relative deviations in percent.

To further investigate how the height of the turbulence wedge influences the wedge detection algorithm, a thermogram of the size (140×140) pixels^2 with a single turbulence wedge with CNR $= 11$ is investigated, and the height $\tilde{h}$ of the wedge was changed systematically from $\tilde{h} = 0.1 \cdot h_{\max}$ to $\tilde{h} = h_{\max}$ in 15 steps. For each wedge height, $n = 50$ thermograms were investigated. The width of the wedge was drawn anew for each thermogram from $\mathcal{N}(\mu = \tilde{h}/3, \sigma^2 = 0.2)$. In Figure 12a, $\langle F_N \rangle$ is shown, while Figure 12b shows the position error, and Figure 12c shows the relative deviations of the wedge's size and area $\langle F_h \rangle$, $\langle F_w \rangle$ and $\langle F_A \rangle$ over the relative true wedge heights $\tilde{h}/h_{\max}$. The gray region in each figure indicates relative heights commonly found in real thermograms, adapted from Table 1.

In Figure 12a, $\langle F_N \rangle$ is plotted over the relative wedge heights $\tilde{h}/h_{\max}$. For a relative height $\tilde{h}/h_{\max} < 0.2$, no wedge is detected by the algorithm because it is too small. Therefore, no information about the positions or the areas is available and the thermograms are excluded from the analysis for the determination of the position error and the relative deviation of the size and area of the wedge. For large wedges near $\tilde{h}/h_{\max} = 1$, $\langle F_N \rangle$ increases to 11%, which means that more than the correct number of wedges are found. This can be explained by the fact that the wedge templates are drawn from an interval of $[0.5, 0.95] \cdot h_{\max}$. When the templates are consistently smaller than the turbulence wedges, two maxima form in the cross-correlation curve close to the actual position of the wedge: One at the left edge of the wedge, one at the right, which leads to the algorithm detecting two smaller wedges close together instead of one large wedge.

Figure 12b shows the mean position error $\langle \frac{\Delta x}{\tilde{w}} \rangle$ over the relative height. Note that the mean wedge width $\langle \tilde{w} \rangle$ was calculated for each height parameter instead of averaging over all. For small wedges with $\tilde{h}/h_{\max} < 0.4$, the largest position error of $\langle \frac{\Delta x}{\tilde{w}} \rangle = 10\%$ can be found. However, for wedges with typical heights found in real thermograms (see gray region in Figure 12b), the position error is only around 1%. For large wedges near $\tilde{h}/h_{\max} = 1$, where more than one wedge is detected, the average position error only includes the position of the first found wedge. The position error therefore drops to $\langle \frac{\Delta x}{\tilde{w}} \rangle = -5\%$ for $\tilde{h}/h_{\max} = 1$, indicating that the first wedge is found slightly left of the actual position. This issue could be solved by using larger templates; however, this wedge size seldom occurs in real thermograms.

In Figure 12c, the relative deviation of the height, the width and the area of the wedge is shown over the true relative wedge height. For wedges with $0.2 < \tilde{h}/h_{\max} < 0.3$, the algorithm determines larger-than-true areas, which is evident from $\langle F_A \rangle > 0$. For wedges larger than $\tilde{h}/h_{\max} > 0.6$, the areas found are consistently smaller than the true areas, which is also apparent in the verification results in Table 2. This can be explained by the intensity gradient in the thermogram, which reaches the intensity value of the wedge at $\tilde{h}/h_{\max} = 0.73$. Therefore, the tips of the wedges in this height range have a CNR close to zero, which interferes with the determination of the wedges' areas.

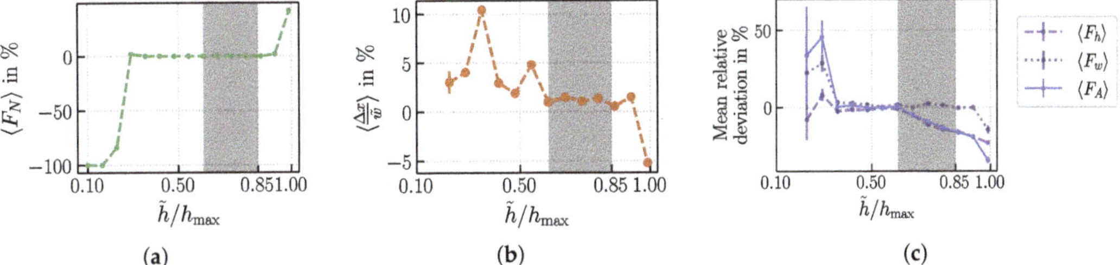

Figure 12. Analysis of the deviation of the height h, the width w and the area A of one wedge in simulated thermogram, averaged over $n = 50$ trials for each relative height value $\tilde{h}/h_{\max}$. The CNR stays constant at a value of 11. (**a**) shows the mean relative deviation of the number of detected wedges $\langle F_N \rangle$ over the relative true wedge heights $\tilde{h}/h_{\max}$. The mean normalized position error $\langle \frac{\Delta x}{\tilde{w}} \rangle$ over the true relative heights $\tilde{h}/h_{\max}$ of the wedges is shown in (**b**). The relative deviations of the height, the width and the area are shown in (**c**). The grey region indicates relative heights commonly found in real thermograms, i.e., $0.73 \pm 1\,\sigma$, with $1\,\sigma = 0.12$. The error bar indicates the standard deviation of the mean.

4.3. Validation: Application on Thermograms of an In-Service Wind Turbine

To validate the MBA, it was applied to real thermograms of rotor blades of in-service wind turbines. In Figure 13a, such a real thermogram of a wind turbine blade with five well-separated turbulence wedges with a CNR value of 6 or more is shown. Since the ground truth is not known, the measured areas can only be compared with a manually created reference, which is depicted in Figure 13b. The angle of the turbulence wedges is specified as 5°. The state-of-the-art results with the GBA are shown in Figure 13c for comparison; the results of the novel MBA are shown in Figure 13d. The GBA only detects the turbulence wedges with CNR > 15, resulting in only two detected turbulence wedges, while the MBA detects all five turbulence wedges, regardless of the CNR value. In Table 3, the values of the determined area, the reference area and the relative deviation of the total area F_A are displayed. Note that the deviation of the total area for the GBA results in $F_A = -64.2\%$, while the MBA only has a deviation of $F_A = -7.7\%$. The deviation value is in agreement with the deviation determined in the characterization of the algorithm in Section 4.2, where for CNR $\geq$ 15, a deviation of $F_A \leq 8\%$ is expected, see Figure 11c.

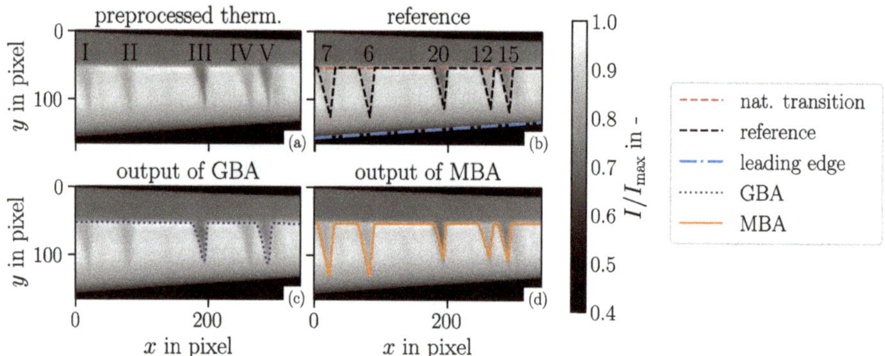

Figure 13. Thermogram of a segment of a rotor blade of an in-service wind turbine with $N = 5$ turbulence wedges. (**a**) shows the preprocessed thermogram with the turbulence wedges numbered with roman numerals. (**b**) shows the reference transition line, which was estimated manually, as well as the CNR of the wedges noted above the respective wedge. (**c**) shows the result of applying the gradient-based algorithm (GBA) of Gleichauf et al. [14] to the image and (**d**) shows the result of the application of the model-based algorithm (MBA) on the thermogram. Note the reduced color scale.

Table 3. Areas determined by the model-based algorithm (MBA) as well as the state-of-the-art, gradient-based algorithm (GBA) for all turbulence wedges in Figure 13. The area values are given in unit pixel^2 and are summed over all turbulence wedges. N is the found or true number of wedges in the image.

	Total Wedge Area A	F_A in %	N
Reference Value	3814.5	0	5
GBA	1364.9	−64.2	2
MBA	3520.5	−7.7	5

In contrast to Figure 13, Figure 14a shows a wind turbine blade with nine smaller turbulence wedges that are close together or even touching. The skew of the turbulence wedges is also larger, so that the angle of the turbulence wedges amounts to 12°. Furthermore, all turbulence wedges have a CNR value below 7. Between the seventh and eighth wedges, a small region with lower intensity might be visible that has a CNR of only CNR = 0.2, and was therefore excluded from the manually drawn reference line, seen in Figure 14b. The GBA again only detects turbulence wedges with CNR > 3, while the MBA again detects all turbulence wedges, see Figure 14c,d. This means that the GBA can detect wedges with a lower CNR in this thermogram than in the thermogram depicted in Figure 13. This is probably due to the preprocessing of the GBA, which applies a histogram equalization to each investigated thermogram. Therefore, different thermograms can be optimized to different degrees, leading to differences between thermograms with regard to the CNR of still detectable turbulence wedges. The MBA does not exhibit this fluctuation in the performance as no preprocessing is needed, which makes the MBA more reliable.

In Table 4, the summed areas of all wedges as determined by each algorithm and the manually estimated reference area are listed. Even though the GBA does not detect every turbulence wedge, the found areas of each wedge are determined to be much larger than they actually are due to the image noise, which leads to a remaining relative deviation of the total area of −5.1%. However, a quantitative comparison of the determined area of each wedge to the true area would show the error in the area determination more clearly. On the other hand, the MBA detects ten wedges, which includes, in addition to the nine referenced wedges, the turbulence wedge with CNR = 0.2 between wedges VII and VIII. However, the determined areas are always slightly smaller than the true areas of the wedges, which in total leads to a relative deviation of −20.2%.

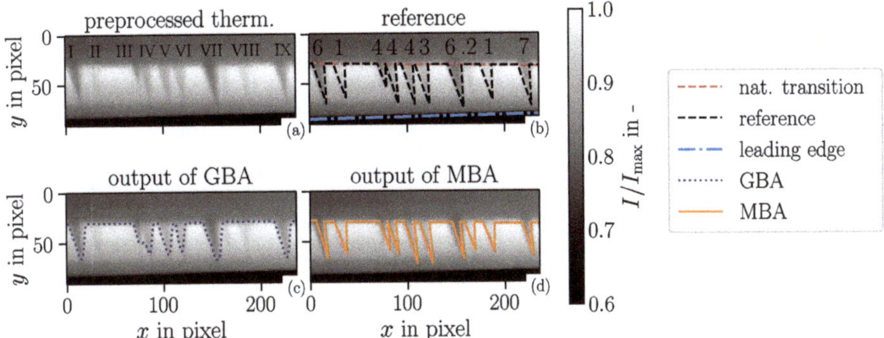

Figure 14. Thermogram of rotor blade of an in-service wind turbine with $N = 9$ turbulence wedges. (**a**) shows the preprocessed thermogram with the turbulence wedges numbered with roman numerals. (**b**) shows the reference transition line, which was estimated manually, as well as the CNR of the wedges noted above the respective wedge. (**c**) shows the result of applying the gradient-based algorithm (GBA) of Gleichauf et al. [14] to the image and (**d**) shows the result of the application of the model-based algorithm (MBA) on the thermogram. Note the reduced color scale, which makes the turbulence wedges more visible.

Table 4. Areas determined by the model-based algorithm (MBA) as well as the state-of-the-art, gradient-based algorithm (GBA) for all turbulence wedges in Figure 13. The area values are given in unit pixel^2.

Area Type	Total Wedge Area A	F_A in %	N
Reference Value	2080.0	0	9
GBA	1972.9	−5.1	6
MBA	1659.0	−20.2	10

The validation demonstrates that the proposed model-based wedge detection algorithm is capable of detecting turbulence wedges and their areas on real thermograms of wind turbines in operation. The algorithm was able to determine wedges with CNR values in a range of $0.2 \leq$ CNR ≤ 20. The state-of-the-art algorithm consistently detects fewer wedges than actually present in the thermogram due to low CNR values, but overestimated the areas for each found wedge. While the model-based algorithm detected the number of wedges correctly, it underestimated the area by 7% to 20% depending on the CNR of the turbulence wedges, which is in agreement with results from the characterization of the algorithm.

5. Conclusions and Outlook

A model-based image processing algorithm for the detection of turbulence wedges on rotor blades of wind turbines in operation was introduced. The algorithm utilizes an image-processing approach called template matching, a technique for finding parts of a thermogram which match a template image. The templates are wedge-shaped, which imitates the natural shape of a turbulence wedge. Using the template-matching method, turbulence wedges are detected as a whole, including position and size parameters, instead of identifying single positions of the premature transition line, which is the case in the state-of-the-art gradient-based method. Therefore, the approach simplifies the determination of the additional turbulent area on the rotor blade due to turbulence wedges, once the turbulence wedges are identified and located. The positions of the turbulence wedges between the natural transition line and the leading edge are detected first using a cross-correlation of the wedge templates and the thermogram. Then, the areas are determined by comparing the detected wedges to wedge templates of different sizes, where the size of the template with the highest similarity measure is finally considered as the measured size of the turbulence wedge.

The model-based algorithm was verified and compared to the state-of-the-art algorithm by evaluating a simulated thermogram with three turbulence wedges of different CNR values. The state-of-the-art algorithm was only able to detect one of the three wedges, leading to a relative deviation of the area of −38.5%. The area of the single detected wedge was overestimated, which is why the relative deviation is not smaller. On the other hand, the model-based algorithm found all three and also estimated their areas with a relative deviation of 1.5%.

A detailed characterization of the algorithm was performed at different CNR values and different wedge sizes. For $4 \leq CNR \leq 16$, which are CNR values commonly found in real thermograms, the absolute deviation of the determined position stayed below 2.5% of the average wedge width. The relative deviation of the wedge areas remains below 10% for $4 \leq CNR \leq 16$. Overall, wedges with a CNR larger than 2 were detectable using the model-based algorithm.

To finally validate the model-based algorithm, it was applied to two real thermograms of rotor blades of wind turbines in operation. A comparison with the state-of-the-art algorithm shows an improvement in the number of wedges detected: the state-of-the-art algorithm consistently found fewer wedges than truly present in the thermogram, leading to relative deviations of the area of up to −64.2%. However, the area of the detected wedges was often overestimated. The model-based algorithm found all wedges but in general underestimated their areas. Even so, the model-based algorithm had relative deviations of at most −20.2%, which, in addition, can be reduced by correcting the systematic error that was determined according to the algorithm characterization. As a result, a clear improvement in the detection of turbulence wedges and their areas compared to the state-of-the-art algorithm was achieved. Hence, a more accurate analysis of the impact turbulence wedges have on the efficiency and the annual energy production of the wind turbine is enabled.

As an outlook, the detection of the turbulence wedges' positions still can be improved. This concerns, in particular, turbulence wedges near the edge of the thermogram, which currently have a lower probability of being detected, but which could be solved by padding the thermogram. Additionally, the determination of the area can be improved. Instead of using linearly decreasing weights for the weighted correlation evaluation, the actual intensity gradient in the thermogram (in particular at the wedge tip) needs to be automatically recognized and then taken into account in the weighting. Furthermore, the wedge detection algorithm can be enhanced to work with subpixel accuracy by using interpolated thermograms and templates. Lastly, an automatic determination of the wedge angle is desirable in the future so that the algorithm then works completely automatically. A further next step is the application of the algorithm for intensive field measurements of wind turbines in motion to investigate the additional area with turbulent flow due to the turbulence wedges over a one year operation and longer. On a shorter time scale, the algorithm will be used to study dynamic flow effects and how they influence the turbulence wedges.

Author Contributions: Conceptualization, A.-M.P., M.S. and A.F.; methodology, A.-M.P., D.G. and A.F. ; software, A.-M.P.; formal analysis, A.-M.P.; investigation, D.G.; writing—original draft preparation, A.-M.P.; writing—review and editing, A.-M.P., D.G., M.S. and A.F.; visualization, A.-M.P.; supervision, A.F. and M.S.; project administration, M.S.; funding acquisition, D.G., M.S. and A.F. All authors have read and agreed to the published version of the manuscript.

Funding: This research was funded by the German Federal Ministry for Economic Affairs and Energy (BMWi) within the project of PreciWind, grant number 03EE3013A.

Conflicts of Interest: The authors declare no conflict of interest.

References

1. Keegan, M.H.; Nash, D.; Stack, M. On erosion issues associated with the leading edge of wind turbine blades. *J. Phys. D Appl. Phys.* **2013**, *46*, 383001. [CrossRef]
2. Corten, G.P.; Veldkamp, H.F. Insects can halve wind-turbine power. *Nature* **2001**, *412*, 41–42. [CrossRef] [PubMed]
3. Traphan, D.; Herráez, I.; Meinlschmidt, P.; Schlüter, F.; Peinke, J.; Gülker, G. Remote surface damage detection on rotor blades of operating wind turbines by means of infrared thermography. *Wind. Energy Sci.* **2018**, *3*, 639–650. [CrossRef]
4. Latoufis, K.; Riziotis, V.; Voutsinas, S.; Hatziargyriou, N. Effects of Leading Edge Erosion on the Power Performance and Acoustic Noise Emissions of Locally Manufactured Small Wind Turbine Blades. *J. Phys. Conf. Ser.* **2019**, *1222*, 012010. [CrossRef]
5. Kusnick, J.; Adams, D.E.; Griffith, D.T. Wind turbine rotor imbalance detection using nacelle and blade measurements. *Wind Energy* **2015**, *18*, 267–276. [CrossRef]
6. Han, W.; Kim, J.; Kim, B. Effects of contamination and erosion at the leading edge of blade tip airfoils on the annual energy production of wind turbines. *Renew. Energy* **2018**, *115*, 817–823. [CrossRef]
7. Carlomagno, G.M.; Cardone, G. Infrared thermography for convective heat transfer measurements. *Exp. Fluids* **2010**, *49*, 1187–1218. [CrossRef]
8. Dollinger, C.; Sorg, M.; Balaresque, N.; Fischer, A. Measurement uncertainty of IR thermographic flow visualization measurements for transition detection on wind turbines in operation. *Exp. Therm. Fluid Sci.* **2018**, *97*, 279–289. [CrossRef]
9. Lang, W.; Gardner, A.D.; Mariappan, S.; Klein, C.; Raffel, M. Boundary-layer transition on a rotor blade measured by temperature-sensitive paint, thermal imaging and image derotation. *Exp. Fluids* **2015**, *56*, 1–14. [CrossRef]
10. Mertens, C.; Wolf, C.C.; Gardner, A.D.; Schrijer, F.; van Oudheusden, B. Advanced infrared thermography data analysis for unsteady boundary layer transition detection. *Meas. Sci. Technol.* **2019**, *31*, 015301. [CrossRef]
11. Crawford, B.K.; Duncan, G.T.; West, D.E.; Saric, W.S. Robust, automated processing of IR thermography for quantitative boundary-layer transition measurements. *Exp. Fluids* **2015**, *56*, 1–11. [CrossRef]
12. Richter, K.; Schülein, E. Boundary-layer transition measurements on hovering helicopter rotors by infrared thermography. *Exp. Fluids* **2014**, *55*, 1–13. [CrossRef]
13. Dollinger, C.; Balaresque, N.; Gaudern, N.; Gleichauf, D.; Sorg, M.; Fischer, A. IR thermographic flow visualization for the quantification of boundary layer flow disturbances due to the leading edge condition. *Renew. Energy* **2019**, *138*, 709–721. [CrossRef]
14. Gleichauf, D.; Sorg, M.; Fischer, A. Contactless Localization of Premature Laminar–Turbulent Flow Transitions on Wind Turbine Rotor Blades in Operation. *Appl. Sci.* **2020**, *10*, 6552. [CrossRef]
15. Vollmer, M.; Möllmann, K.P. *Infrared Thermal Imaging: Fundamentals, Research and Applications*; John Wiley & Sons: Hoboken, NJ, USA, 2017.
16. Abràmoff, M.D.; Magalhães, P.J.; Ram, S.J. Image processing with ImageJ. *Biophotonics Int.* **2004**, *11*, 36–42.

MDPI

St. Alban-Anlage 66

4052 Basel

Switzerland

Tel. +41 61 683 77 34

Fax +41 61 302 89 18

www.mdpi.com

Applied Sciences Editorial Office

E-mail: applsci@mdpi.com

www.mdpi.com/journal/applsci

www.ingramcontent.com/pod-product-compliance
Lightning Source LLC
LaVergne TN
LVHW070949170726
843515LV00005B/946